The Right Light

The Right Light

Interviews with Contemporary Lighting Designers

Nick Moran

First published 2017 by
PALGRAVE

Palgrave in the UK is an imprint of Macmillan Publishers Limited, registered in England, company number 785998, of 4 Crinan Street, London, N1 9XW.

Palgrave Macmillan in the US is a division of St Martin's Press LLC, 175 Fifth Avenue, New York, NY 10010.

Palgrave is a global imprint of the above companies and is represented throughout the world.

Palgrave® and Macmillan® are registered trademarks in the United States, the United Kingdom, Europe and other countries.

ISBN 978–1–137–33478–7 hardback
ISBN 978–1–137–33477–0 paperback

This book is printed on paper suitable for recycling and made from fully managed and sustained forest sources. Logging, pulping and manufacturing processes are expected to conform to the environmental regulations of the country of origin.

A catalogue record for this book is available from the British Library.

A catalog record for this book is available from the Library of Congress.

Contents

Acknowledgements and Thanks

Without the generous participation of the following lighting designers, this book would not exist (website URLs correct at time of writing)

Neil Austin (www.neilaustin.com)
Lucy Carter (www.loesjesanders.com/lsl/clients/cvs/carter_cv.html)
Jon Clark (www.jon-clark.com)
Natasha Chivers (natashachivers.co.uk)
Paule Constable (www.pauleconstable.com)
James Farncombe (www.jamesfarncombe.com)
Rick Fisher (www.ald.org.uk/rickfisher)
Mark Henderson (www.markhendersonlightingdesign.com)
David Howe (www.dhld.biz)
Michael Hulls (www.michaelhulls.com)
Mark Jonathan (www.markjonathan.com)
Peter Mumford (www.petermumford.info)
Ben Ormerod (www.benormerod.com)
Bruno Poet (www.brunopoet.co.uk)
Paul Pyant (www.ald.org.uk/paulpyant)
Nick Richings (www.nickrichings.com)
Johanna Town (www.johannatown.co.uk)
Hugh Vanstone (www.hughvanstone.com)
Katharine Williams (www.katharinewilliams.co.uk)

Margaret Bowden, Emma McGrath and UK Transcription Ltd for transcribing the interviews.

Colleagues at Royal Central School of Speech and Drama and on the executive of the Association of Lighting Designers for advice and support in developing the project.

Editors and others at Palgrave for their help, advice and support.

And my life partner Shelagh Prosser for understanding and support.

Royalties from the sale of this book are being donated to the charity Light Relief, a charity established by and for the entertainment lighting industry, to help in times of extreme hardship.

1 Active Practice

What makes the light *right* on stage? Of all the almost numberless choices available, why does the lighting designer (LD) choose this one over all the others? That is the question at the heart of the 19 interviews with lighting designers on which this book is based. By *right light* I'm not arguing for an absolute – a single perfect solution to each production's lighting design. Rather I'm proposing *right* as a continuum, each solution requiring to be judged more or less right, from only just right and right enough for now (but requiring work if time permits) through right for this moment (but not that one) all the way to so right I can't imagine anything improving it. And throughout we will naturally be looking at light in the context of many other elements that make up the stage picture, including the set, costume, projection and sound design, direction and choreography, performers and audience.

Through the interviews, I'm hoping to reveal something of the working process of some of the top lighting designers in UK theatre today; their creative starting points, their priorities and the basis of their aesthetic choices, their triumphs and insecurities, and what they think of as good lighting design. I'll also be asking the question: is lighting design for live performance 'art'?

What singles out the LDs I wanted to interview is that it seems to me that their approach to lighting the stage is quite different to what has gone before, and to some extent what continues to happen in many theatres. This book suggests that these practitioners are all part of a new way of doing theatre lighting design, responding to changes in the way theatre – drama, opera and dance in particular – is both made and presented.

How to use the book

Much has been written on the technical aspects of lighting the stage – the 'how to' of lighting design. Much less has been written concerning the 'why?' This book aims to correct that. It aims to provide a non-technical insight into the process for readers working in or interested in theatre and performance that is also useful for students of lighting design and aspiring lighting designers.

I have tried to avoid technical details whenever possible and to focus on the creative side of lighting design practice. However, interviews frequently became conversations between two 'experts' with similar experiences – so there are times when quotes from interviews have required some words to be inserted to provide context for the more general reader. These are in [square brackets]. There are also a number of specialist words and phrases and instances where words are used in different ways to common speech. I hope that most of these words can be found in the glossary at the end of the book. In the interviews, I've used a short dash (–) to indicate a pause and three dots (...) to indicate that text has been cut out. I've kept in a lot of colloquial use in the hope that the reader will get a closer connection to each lighting designer, but with the permission of the interviewees I have tidied up some grammar to make meanings clearer.

There are no pictures in this book. With a few exceptions I believe that the subtlety and strength of most theatre lighting design is poorly served by still photographic images in books. One or two still images out of context do not really tell us very much about the light of a production. However, good-quality images of almost all the productions referred to at length can be found on the internet, frequently with short video clips. To make it easier to find these is one reason why I have included so much production information in the notes. However, it must be said that cameras do not see the world in the same way that humans do, and the only way to really experience the lighting design for live performance is to see it live – something I hope that readers will do.

The beginnings of the role of lighting designer

Theatre as an art form is at least 3000 years old. Theatre lighting design in the way it is practised in North America and much of Europe today is probably less than 100 years old. Theatre lighting design in the UK has a heritage, even if it is not a very long one. Some of today's top UK lighting designers began their practice watching the work of, or working for, the first generation of people who were called theatre lighting designers.

Once theatre went indoors – in Europe this was largely during the Renaissance – it became almost essential to provide some kind of artificial light to enable the audience to see the performers. Theatre has a long-standing love affair with technology, from the architecture and masks of classical Greece, through the fantastical stage machinery of Renaissance Italy to the digital projection technologies so prominent today. Although some theatre makers have at times sought to purify their art by removing technology from their stages, its advance is at least a part of the story behind most advances in theatre-making practice.[1] Over the long history of theatre, lots of people have come up with ingenious ways of getting light onto the stage, and controlling it too. The introduction of gas lighting into theatre was seen as very much a mixed blessing: more light on stage for sure, and better control of it, but also more heat, and a noxious smell. For theatre, as for much of Western society, one of the most important emerging technologies at the end of the nineteenth century was electric light.

Although gas lighting had considerably extended the area of the stage on which the performers could be clearly seen, in most places it was the introduction of controllable electric lighting that finally enabled significant changes to be made in the ways drama is performed and theatre is presented indoors. Once controllable electric light became established in theatres, it became possible and more usual to dim the lights in the auditorium during the performance. The increased flexibility of electric light meant that more of the stage could be adequately illuminated, and performers could move more freely about the stage. Many other radical changes were taking place at this time, in theatres and in the wider societies they were part of, and these too had profound influences on the stage. However, it is hard for me to see how the move from the declamatory theatre style of nineteenth-century theatre towards more naturalistic styles of performance could have developed on a stage lit by gas footlights and limelight follow-spots.

As part of this wider revolution, the roles of director and stage designer as we know them today began to become the norm. Alongside these new *creative* roles, new *technical* roles appeared too. In the lighting world, the specialist knowledge required to master the technologies of the new electric theatre lighting required a chief electrician (master electrician in North America), usually heading a technically focused team. The job title on both sides of the Atlantic reflects the main responsibility of the role then (and in some places now) – that is, the electrical system that powers and controls the lighting instruments, rather than the qualities of the light on stage – what we now call the lighting design. These electricians were generally given instructions as to what to do with their lighting system by the producer/director or sometimes by the stage designer.

Quite early in the process of bringing controllable electric light onto the stage, directors and producers began to acknowledge the potential of this new medium. By 1925 C. Harold Rudge felt able to write in his book *Stage Lighting for 'Little' Theatres*:

> A play may be good, and it may be well acted, but it will fail unless the audience can see it. Light therefore plays a most important part in

> the theatre. The first duty of the electrician [sic] is to make the actor visible to the audience; his [sic] second is to aid the action and atmosphere of the play by doing this in a suitable manner. ... [G]ood acting can only be enhanced by suitable scenery and beautiful lighting (Rudge, 1925, p. 71).

In many ways, the fundamentals as expressed here have not changed. All the LDs interviewed regard the appropriate illumination of the performers to be the priority – most of the time. They also agree that the next priority – most of the time – is to aid the action and atmosphere, and that their work is there to enhance the performance. Rudge is, however, a little disingenuous when he implies that he will leave it to his electrician to decide how the light should look on stage. (Rudge does not use the word 'design' in relation to light or lighting.) Later on in the same book he writes:

> In poetical plays, or any play that is not mounted realistically, the producer can proceed boldly and unhesitatingly with the lighting (Rudge, 1925, p. 76).

The general expectation is that the producer (adopting a role that would later be called director) will be the one doing the job we would now call 'designing the lighting,' while the electrician does what he [sic] is told.

The job of lighting designer became established in the USA at least a generation earlier than in the UK. The US stage lighting pioneer Stanley McCandless – who is also credited with starting academic study and training in lighting for the stage – first published his influential work *A Method of Lighting for the Stage* in 1932. In his introduction to the revised edition (published in 1939), he writes that his purpose in writing the book is,

> to give the young designer or technician the confidence with which to face the real problems of lighting. The art of illumination is not measured by ingenuity, although the complicated technical nature of the subject often leads people to applaud technical mastery ...
>
> This plan prepares the palette, as it were, of the lighting designer, and suggests a practical method of using the tools that are available, but it does not pretend to guarantee the final results of balance and composition in dramatic pictures. The final result depends on the eye and taste of the designer (McCandless, 1932 Revised edition 1939, pp. 9, 10).

So here we see that the job of lighting the stage has begun to be associated with design *and* the job is no longer assumed to be the role of the producer/director. Today, McCandless is frequently accused of proposing a formulaic craft-based approach to stage lighting (usually by people who have not actually read his work), but what is quite clear here is that he believes an artistic approach is required to light the stage well – to choose the *right light* for each moment of the performance.

By the late 1940s in the UK, perhaps as part of the post-war celebration of the democratic spirit, it had become usual to acknowledge set designer and costume designer in theatre programmes. By this time too, the role of theatre director was beginning to be understood as the person in charge of a *team* of theatre artists, rather than a solo authoritarian creator. However, in 1956 Geoffrey Ost in his handbook *Stage Lighting* (complete with a foreword by the great actor/manager Donald Wolfit) was still able to write:

> Readers will soon realise that the electrical and technical side of the business as only a means to an end is not difficult to grasp, and that the more important part of the work is arranging and directing the light on stage.

Still no mention of design. Mr Ost goes on:

> In the early stages of play production he (the producer, for whom this book is primarily intended) carries in his head a mental impression of the play as it will eventually appear. Therefore, it is highly desirable that he should plan his own lighting (Ost, 1956, second impression 1957, p. 11).

So this was the general expectation in the UK as the likes of Michael Northen, Robert Bryan and the legendary Richard Pilbrow were beginning their careers. Many directors and some set designers still expected to 'light' their own shows. Slowly at first and only in a few places, the role of the specialist lighting designer as an artistic collaborator became established.

By the time the LDs in this book began their practice, the theatre lighting designer had become a regular member of the team working together to make a *show*. For some there remained the expectation that the lighting designer was there mostly for their technical expertise rather than their creative input.[2] However, all the LDs interviewed for this book (and many others too) have largely escaped that way of working.

Today it is common practice in the UK and elsewhere to refer to the 'creative team', which will include designers (set, costume, sound, projection), other specialists (choreographer, musical director/conductor, etc.) and for new work, the writer(s) and/or composer(s). Almost always the acknowledged head of the team is the director. This is now the generally accepted structure for making work for the stage in North America and Britain, and increasingly elsewhere too.[3]

The new kind of practice

The lighting designers interviewed here are among a group that span several generations, but who are all to a greater or lesser extent working in a way that is different to the first generation of UK-based lighting designers, and many still working in the UK today. Some of those interviewed trace their early influences back to the first people who were credited as lighting designers in the UK. One of the most influential of these remains Robert Bryan – universally known as Bob. He is acknowledged by Paul Pyant and Mark Jonathan as having had a major influence on their way of thinking about light, particularly in opera. Nick Richings worked closely with the late Michael Northen, who is often cited as the first person to be billed as lighting designer in the UK. The links back to these first practitioners continue through to the next generation. Ben Ormerod, who acknowledges his debt to Gerry Jenkinson, a near-contemporary of Bob Bryan, is in turn an inspiration to, among others, Bruno Poet, Neil Austin and Paule Constable, all of whom have worked as Ormerod's associates, while John Clark has in turn been associate to Paule Constable.

To be clear, what I'm writing about is the practice of some of the most creative lighting designers working in UK theatre at the moment. This is where I live, teach and research, and it is not my intention to generalise beyond UK theatre. Although most if not all the LDs interviewed in this book work internationally, they are all based in the UK, and most of the work mentioned was produced here too.

I'm going to call this newer approach to creating work 'active practice', correlating it with another relatively new concept of active aesthetic for which I am grateful to my colleague at Royal Central School of Speech and Drama, Dr Experience Bryon.[4] The active aesthetic concerns itself with the *way* of practice rather than the *what*. It sees practice as a dynamic, carrying within it, a sense of responsibility and ability to respond to the ways in which we engage in creative acts. My *active practice* is contrasted to the traditional practice that has been more or less the norm for much of the second half of the twentieth century. By aligning it with *active aesthetic*, active practice can open up a way of doing things that involves being inside a process as opposed to essentially responding to the creative acts of others.

For many theatre makers working in all kinds of genre today, it is no longer enough for light to be just the final layer added to an almost completed piece. Instead they aim for light to be an integral part of the development of the work, its role within the piece considered from as early in the production process as possible, and this requires the LD to be an active practitioner. I think that for many of the LDs here, when I refer to them as active practitioners they are inside the creative process, and that fundamentally affects the way in which they are *doing* lighting design. What informs the ways in which they are making choices is the *active aesthetic*, and this in turn informs what they mean by the *right* light.

Following Bryon, then, this concern with the active aesthetic is a marker of the integrative lighting designer.

It feels to me as if in this second decade of the twenty-first century, theatre lighting design, in the UK at least, has come of age. Few in UK theatre now question the need to have a competent lighting designer involved in almost any theatre production. The expectation is generally that this person will be an active creative partner in the production team. This is in contrast to the general expectation of previous generations of theatre makers who, if they had an LD at all, saw him [sic] as primarily the leader of a team of technicians who turn the lights on and off as and when the director tells them. This is not to say that every lighting designer of previous generations simply did what the director told them. It is only because many of them developed an individual creative practice, which showed directors and others what was possible, that the present generations have the opportunities to take this further.

In the context of this discussion, a theatre production might be a play, an opera or musical, a ballet or another form of dance piece, or something that is not quite any of these. Today in the UK there is an expectation not only that designed light will play a part in helping the piece live, but also that a lighting designer will be responsible for designing that light. It is important to remember that this has only recently become the norm, and is by no means universally true outside the UK and North America.

Lighting design is frequently mentioned in newspaper and online theatre reviews and even in comment pieces in some more serious newspapers. Most of the major theatre awards ceremonies honour lighting designers. Describing theatre lighting design as an *art* and theatre lighting designers as *artists* no longer causes raised eyebrows among the 'great and the good' of UK theatre. Today the practice is mature, and what both audiences and theatre professionals expect of light on stage is far more than the visibility and atmosphere of traditional practice. Increasingly there is an expectation that the lighting on stage will play a significant role in telling the story, and that lighting designers will be concerned with much more than the technical realisation of a decorative aesthetic.

How then might critics, theatre academics and other audience members recognise this different approach to theatre lighting design? A lot of the time the most visible aspect of this approach can be characterised as 'less is more.' On stage, there is now a much greater willingness – some would say demand – for light to be used *dramatically* – as a spatial design tool and a signifier, for example. The notion of doing this goes back at least as far as the early-twentieth-century writings of Adolphe Appia and Edward Gordon Craig. Both men wrote about the power of light on stage to shape space and to be a symbol.[5] The bold and conscious use of light in this way has for a long time now been integral to the performance of some kinds of contemporary dance, and was often a feature in opera houses, but was not really seen on the drama stage in the UK until relatively recently. (There is some irony in reporting that for much of my early working life I was frequently cautioned to avoid what was generally called 'dramatic lighting' on the drama stage, as it was thought that light should not 'draw attention to itself'.)

On the drama stage, the main objective of the lighting design has been, and remains, the visibility of the performers. Richard Pilbrow's influential 1997 book *Stage Lighting Design* includes the lines:

> The cardinal rule is: Each member of the audience must be able to see clearly and correctly those things that he is intended to see. ...
>
> Ninety-nine percent of the time it is the designer's duty to light the actors clearly so that everyone can see them. (Pilbrow, 1997, p. 7)

For all the lighting designers I interviewed, visibility of performers on stage remains key, and getting light into eyes is especially important. Here is Ben Ormerod, whose influence on several of the other LDs interviewed here has already been noted, talking passionately about the importance of getting light into the actors' eyes:

> **Ben Ormerod:** [Visibility of the actors] is absolutely key. It is the most important part of the job. If you rig a light that doesn't get into someone's eyes, you've got to justify its existence on stage. Even backlights can be designed to get into people's eyes. Every light has to earn its keep. If you're a young lighting designer, starting out, and you're lighting a show with 20 lights; if one of those lights doesn't light someone's eyes, what is it doing there? It's as simple as that.

I'm confident that when talking about drama or opera, no one interviewed here would disagree with Ben on this. And yet, by 'visibility' these LDs don't necessarily mean the same as Pilbrow did when he wrote the quote cited earlier. In the interviews they argue, in different ways, that while there will be many moments when 'lighting the actors clearly so that everyone can see them' is key, there are approaches to achieving this that are far more interesting and useful (right?) than the carefully planned and focused 45-degree washes of McCandless and his followers that dominated traditional stage lighting practice for at least two generations.

Mark Henderson, who has more awards for theatre lighting than any other UK lighting designer and perhaps grew up in that tradition, had this to say on the subject:

> **Moran:** Getting back to angles – through doing this project, I've realised that my process has changed markedly, in that I used to start by putting in a 45/45-degree front wash and a backlight bar – and I just don't now.
>
> **Mark Henderson:** Oh yes – that's what I used to do, and absolutely don't now.
>
> **Moran:** What do you think caused that change, because that approach has almost disappeared in this country?
>
> **Mark Henderson:** Yes – I think it leads to more dramatic looks – you don't get a dramatic look with a 45/45 [degree front cover] and straight backlight. You need dramatic angles to create dramatic pictures, and I think that's where it's come from.

Active practice involves a different kind of planning – and frequently a much closer collaboration between LD, director and performer. Alongside this though runs a broader definition of 'light them clearly so everyone can see them.' The traditional practice – perhaps best explained by McCandless, was to ensure the actor's body, and particularly the head, was 'well lit' from at least three directions. Rick Fisher, whose approach is most often to start building a lighting state from the light that will shine on the actors' faces, makes a clear distinction between lighting faces and lighting the whole head:

> **Rick Fisher:** But what it comes back to is we are lighting people's faces. I think, not that I think about this too much, but I think sometimes we make the mistake of thinking that people's faces go 360 degrees around their head. So that means we [try to] light the whole head evenly, but actually the face is only one quarter of the head. I don't need to light the back of their head so much, and I don't necessarily need to light the side of their head so much – I want to light their face.

Rick later talks about paring away surplus light, as do many others interviewed here. The desire is often to do as much as possible with as little light as is practicable – less is more – just the light that is needed and nothing more.

The roots of the active aesthetic

The change in the practice of lighting design that I'm writing about here has, at least in part, been made possible by technical advances. Alongside these though came changes in ideas about what acceptable lighting looked like on stage, influenced by (amongst other things) what was happening on our television screens. Nick Richings and David Howe both argue that the dynamic and bold use of light on television (according to Richings beginning with the American crime lab franchise *CSI*) has helped make the increased use of bold lighting on the drama stage acceptable:

> **Nick Richings:** It would be a very dull play if you just had grey daylight streaming in all the time, because it's not dramatic.
>
> **Moran:** It would look like telly.
>
> **Nick Richings:** It would look like telly before *CSI* came along, or digital cameras. The thing that's changed TV is digital cameras, and then people's realisation that you can make it look like you want it to with light – so you can have big bold highlights, and backlight and key light and all that stuff we only ever used to see in film. I think shows like *CSI* have changed the way we look at everything, including the stage.
>
> That has helped develop a language that most audiences can relate to. You can have someone who is just side-lit [on stage] now. And directors and producers buy into that too, which is the other battle won. You know, I'm quite happy having someone lit in just a shaft of light coming through an open doorway ...
>
> The *CSI* effect has changed people's perception of what's acceptable – and you can do more dramatic things [with light].

David Howe is quite clear about what he and others mean by a 'dramatic' look:

> **David Howe:** What we're looking for [as theatre makers and audiences] is the more sculpted look. We know they have got faces. We know they have got eyes and teeth. We sometimes want to see them, but not all the time. We know they're a big star from the TV. But also we're used to seeing them in the half-light on the TV.
>
> **Moran:** Nick Richings was saying that as well, that he thinks that programmes like *CSI*, and the visual style of that, has influenced what is acceptable, and possible.

> **David Howe:** Yes, absolutely. ... Back in the time of the [1970s TV series] *Upstairs Downstairs,*[6] back then everything was very visible the whole time. If we were doing *Upstairs Downstairs* today [on television] it would be a shadowy basement, with the light coming through a grill [in the wall] over there, and they would be illuminated by the gas jets, or the flames of the fire, or whatever.

Peter Mumford, who is also a film maker in his own right, has this to say on the subject:

> **Moran:** Relatively recently television has become ... unafraid to use light in the same way that the best film makers have done.
>
> **Peter Mumford:** I think that's absolutely true. I think people [now] have a language too, that they don't even know they've got. Through generations of watching film and television, they understand editing without knowing it – for the most part. Obviously there are those who do know it too. But general audiences read a language of editing [on stage that they have learned from the screen]. They read a language of parallel action. They understand flashback and a whole load of things that I imagine an audience of say 60 years ago wouldn't have understood, and wouldn't know how to read. So all of that interacts.
>
> I think now there's a lot to learn from television. In my early days we were all trying to copy film quite a lot – you know HMIs [short-hand for large, powerful film lights] and single shadows to make it look like film. And then discovering that it's a line somewhere between the two, between film lighting and traditional theatre lighting. That is what you actually want. On stage, you can't be as purist as you would be on a film set.

Here Mumford is acknowledging both the debt that contemporary theatre practice of many kinds owes to cinema and television, and the fact that live theatre is different: 'You can't be as purist [with light on stage] as you would be on a film set.'

Something else audiences might have noticed, especially if they sit up at the top of the theatres, is a change in where the light is coming from. I grew up going to see and then working on shows lit largely by ranks of profile spotlights high up on either side of the auditorium. Although still present in many theatres for many of their productions, as Mark Henderson confirmed, this front of house position is no longer the starting point and main tool of the LDs interviewed for this book. (The equivalent positions in the largest two auditoria of London's National Theatre are sometimes referred to by the lighting staff as 'the most expensive call lights in the country' as they are rarely used for anything except the curtain call – the only time when the cast face directly out into the audience.) The LDs interviewed here offered several different reasons for the move away from lighting design built around large front of house rigs. These range from changes in acting and presentational style through to the willingness of technical crews to find alternative hanging positions for them, suggested to me some years ago by Ben Ormerod, and here by one of his prodigies, Neil Austin:

> **Moran:** We were talking earlier about the move from three-quarter front of house lighting to side light ...
>
> **Neil Austin:** [T]hat's been done in dance since the 1950s, earlier even. And it should have made the leap [to the drama stage] much earlier than it did.
>
> I think it comes from two areas: A) electricians who are more interested in rigging in unusual places – so once you can persuade an electrician to rig a whole load of stuff on a boom (which is a real pain in the arse for them), then you can start using it. B) It's the lowest-cost way of colouring a stage with the least amount of lanterns, so it's certainly come from our fringe careers as well – from when we didn't have very many lights. How do you use three lights and cover the entire stage? If you put them above, it will be no use at all but if you put them on one side you could. ...
>
> On top of that what it gives is that super high-lit person within a [darker] environment, and that's what everyone wants and why it gets used a lot.
>
> **Moran:** High contrast between foreground and background?
>
> **Neil Austin:** Yes. It only works if you don't have side walls, but often you persuade the scenic designer to [have the scenic artists] paint the sides down, at least up until head height. Let's have a nice piece of wooden wainscoting please around the set and then you can have whatever colour [light] you want around that.
>
> Actors have become more used to playing in it too.
>
> **Moran:** It does to some extent reflect acting style as well in that less is played straight out, more is played across – a more interior kind of presentation?
>
> **Neil Austin:** Yes. But even in a straight out version you would use the 'Hendie' lighting position. [A lighting position ideally around 2 metres

> above stage height towards the sides of the auditorium, that provides a place to hang lanterns that will light straight into the eyes of actors – named for Mark Henderson, because many of these positions were originally installed to facilitate his lighting designs.] They're called Hendie rails all around this country, and abroad now. I went to LA and put some in the Ahmanson Theatre and told the electrician what I called them and he misheard and called them Handy rails because they're quite handy.

What motivates these LDs to make use of other positions is the strong desire to give three-dimensional shape to the performers in the space. Paule Constable is clear why she avoids using the more traditional 45-degree front of house lighting angle:

> **Paule Constable:** My big obsession is that if you've already made something you are really excited by, and then you use a 45-degree angle from front of house, you turn the light on from out there – and it all just goes... because you're pushing the actors on stage into the floor – and what it is doing so often is NOT making people more vivid, not bringing the people on stage into our lives, but actually it's taking them away from us, and pushing them into the background. ...
>
> 90 percent of the time you're carving something, and when you just add something from front of house – that just completely works against how you're trying to deliver an image, how you're trying to paint – and you kill it!
>
> So in a way, it's by taking everything back to its simplest – this is about light through a window, or this is about this or that light – then when you add, you become very aware of what the repercussions of adding are. So it's trying to find ways to support the text when it is needed but still maintaining a live space that is beautiful, really.

I think the notion that it is part of the LD's role to help to create a connection between the stage and the auditorium is relatively recent too, though the first generation of lighting designers were largely responsible for getting rid of footlights, the biggest barrier between performer and audience. The idea that the LD needs to strive to maintain a 'live space' that 'supports the text' is something that all the LDs interviewed talk about in one way or another.

For Constable and others, the way to achieve this is to work at revealing the three-dimensionality of the performers on the three-dimensional stage. Although she, like many of those interviewed, frequently uses the analogy of painting, here Constable is moving beyond that towards a sculptural aesthetic. This, and the notions of liveness and connectedness of the audience with the lit space, and the focus on revealing the three-dimensionality of performers in that space, are all key components of the aesthetic concerns of the active practice of stage lighting design.

Constable's concern with simplicity too is key to her active aesthetic. As Bryon writes, 'the best choices are not complicated' (ibid):

> **Paule Constable:** I love the simplicity of a single source shadow. That's one of the reasons I love one light doing a great job – it can speak so much. I also love a sense of directional shadow, and I love when the key light is so strong you don't see the shadows of other lights, and all that kind of thing. I love it in its purest form, when the shape that the shadow makes is as interesting as the thing itself. But I think all of those things I love are about that purist sense of it. Because I think I am a bit Amish about all this (laughs)
>
> But I couldn't take an element like shadow and take it out of what I do, but I think it's hugely important for when it says the simple thing that I love to say...

The influence of theatre in the round

Paule Constable and Johanna Town cite the influence of theatre in the round. For Town in particular it is the Royal Exchange theatre in Manchester, which has been an influential producing theatre since its beginnings in 1976, and has always had a concern with light. Town, who trained as a theatre electrician there, was inspired by the productions she worked on and saw at the Royal Exchange, which showed her what was possible. Later, while working with the traditional lighting set-up of a proscenium theatre (with its 45-degree front of house rig and all the other elements of traditional lighting practice) everything looked dull and artificial:

> **Johanna Town:** I just couldn't believe that nobody was making these lovely spaces that were real places that the audience could sort of touch, which is what I'd been brought up with for three years at the Royal Exchange.
>
> [F]or theatre in the round, you need to light on three sides, so you shouldn't have somebody under-lit [but that's not the same as making the light look the same from every side]. What you should be doing is saying the light's coming from over there – a key ... say the action is in a living room and the sun's coming through the window on the right side, the person will need a fill tone on the side away from the window and have

> daylight the side facing the window. That way you have lit someone from both sides but in a more natural way – you have *got to shape them.*
>
> I think I've always taken what I've seen happen in the round and then tried to make it work with a pros arch.

In the 1970s and 80s, many of the theatre spaces producing the most interesting new work were not traditional proscenium theatres. Among these were the Bolton Octagon, Scarborough's Stephen Joseph Theatre and Glasgow's Citizens Theatre – all theatres in the round, and all mentioned by at least one of the LDs interviewed as important starting points for their careers. As if to reinforce the point, the new National Theatre building opened in London in the mid-1970s, and only one of its three theatres had a proscenium. It seems reasonable to me to propose that this preponderance of non-traditional theatre spaces is at least partly responsible for producing, enabling or requiring a radical change in practice – from more than just lighting designers. Significantly though, several LDs can't really be said to have come from this background. Paul Pyant and Mark Jonathan both grew up (from a lighting point of view) on the lighting crew of the Glyndebourne Festival Opera house, a traditional proscenium theatre, and the early work of both David Howe and Nick Richings was predominantly in proscenium theatre.

Ben Ormerod, who was one of the last LDs I interviewed, was keen to point out that many of those interviewed had worked in a range of genres. Almost everyone has a strong background in at least three of: opera, musical theatre, drama and/or dance, and many have also worked on site-specific projects. The successful mixing of scale and genre is perhaps one of the markers of a mature lighting design practice.[7] However, this is not particularly different from past generations of lighting designers.

What is perhaps a marker of the step-change in lighting design practice, however, is that these lighting designers, like Johanna Town, take what they have learnt and developed in smaller spaces with the audience on more than one side, onto large stages for full-scale operas, musicals and dance works, and vice-versa. These lighting designers demonstrate the ability to light intimately in a large space, or grandly in a small space, bringing sensibilities developed lighting dance onto a drama stage, or from drama into a musical theatre production. To quote Jon Clark, relating the advice Paule Constable gave him in preparation for his first design on the immense open Olivier stage of the National Theatre: 'treat it like a Studio theatre and be bold with it. ... If you start piling light into everywhere then you lose any sense of the picture.'

A nice even wash

When I was growing up in theatre, in the 1970s and 80s, on most commercial stages and elsewhere, one of the markers of a good lighting design was that performers looked the same wherever they were on the stage, particularly in the third of the stage closest to the audience. In an interior setting, the presence of a window at one side of stage would be reason enough for light from that side to be warmer or cooler than light from the other side of stage, but generally it would not be sufficient justification for making that side of the stage brighter. For an exterior scene, the high-contrasts light of nature would be sacrificed for the achievement of similar levels and qualities of illumination over the whole acting area. (The single shadow, such as one might see in nature, would be more or less impossible to achieve for technical reasons too.)

The aspiration to create even illumination over the whole acting area dates back at least as far as McCandless, and is perhaps a reaction to candle- and gas-lit stages where this was certainly not the case. Achieving the uniform washes of light that this requires is no mean technical accomplishment, but it is something the first UK lighting designers did very well. However, it often worked against the architecture of the set, and very often it was not dramatic, in the sense that Nick Richings and Mark Henderson mean the word.

These were the even washes that Town talks about being shocked by when she left the Royal Exchange theatre, and theatre in the round contributed to the decline of this style of lighting at least as much as Richings' *CSI* effect did. But so did an aspiration to do more with light on stage, and the gradual rise in the status of the theatre lighting designer.

The integrative lighting designer

The lighting designers here, and many others too, respond much more directly to the architecture of the set, the dynamics of the space, and the performers. They also want to create dramatic lighting that makes beautiful stage pictures.

Integrative lighting designers are more likely to make a researched and thoughtful response to each new production than to impose a style or template on it from outside. They are less inclined to work to a formula expected of this space or that genre. Though they are much more inclined to take what they have learned works in one kind of space, in one genre of performance, and try it out in the new production, that only happens if their research finds that this might be a suitable approach. They walk into each project much less inclined to follow the rules, either their own or someone else's, though they may well develop a set of rules for how light works in each production – a language for the lighting design that works with the other languages of the piece. These rules come from within the text and the production and are not imposed from outside it.

Other changes in the production process

How has all this become possible? Within the production process, away from the general audience and most other critics and observers, other aspects of the shift are evident. The traditional model of practice is for the LD to be brought in only once the director and set designer have developed the underlying concepts that will inform the production, and the space has been fully designed. Many of the lighting designers interviewed here are more regularly part of the discussions about the production from the beginning; that is from the same moment that the set (and costume) designers come on board – occasionally before even that point. Again in the older model, the LD was at the first read-through on day one of rehearsals and then at final run-through in the rehearsal room – which usually happens the day before the beginning of the technical rehearsals in the theatre. Almost all the LDs interviewed aim to spend more time in the rehearsal room earlier in the production process. There is more to be said on this in chapter 2: Instinct as Inspiration.

In the older, traditional model of theatre making, the director is at the top of a hierarchy and suggestions (more often instructions) only flow down. In productions with a flatter structure – where the director is open to extensive two-way collaboration – lighting designers get involved with many aspects of making the piece, and others make suggestions about light too. Several of the lighting designers interviewed recount their involvement in discussions about blocking, character motivation, and even in one case (Neil Austin) a small re-write of the libretto.

Lighting designers need to collaborate with others, too. UK theatre works largely with a deputy stage manager in the role of show caller following the script or score and giving cues to follow-spot operators and someone who is operating a lighting controller (amongst others). When working with larger and more technically complex lighting rigs the LD generally requires a dedicated lighting programmer. When the work is in repertory or on tour, the lighting design will need to be reproduced – generally by a staff member of the repertory theatre or an associate of the original lighting designer. Increased technical complexity has meant that the scope and integrity of lighting design is often dependent on, and at times constrained by, these collaborations.

For the most part the changes in practice here are largely incremental as distinct from the step-change in other areas of lighting design. The exception may be the collaboration with the lighting programmer on the larger productions. These collaborations are part of the focus of chapter 4.

Key issues in conceptualising lighting design

Why is light an important component of stage performance?

Professor Erika Fischer-Lichte, a distinguished international theatre scholar, succinctly explains in her 1992 book *The Semiotics of Theatre*[8]:

> Both natural and artificially produced light – like naturally and artificially produced spaces – can be interpreted in terms of their practical and symbolic function. Making a space visible is, generally speaking, a practical function of light. Illumination of space is what first allows the latter to become evident ...
>
> In addition to this basic practical function – and usually as a consequence of it – light can assume a wide variety of symbolic functions ... Nearly every culture not only has learned to interpret light as a sign in relation to the time of day and season of the year, but has also developed a rich store of light symbolism ... (cited in Palmer, 2013, p. 70).

Fischer-Lichte here is following any number of writers on theatre who remind us that anything that is put on stage, in addition to its practical function, is available to the audience to be interpreted. It seems natural for us to look for layers of meaning in most forms of story-telling, and theatre is no exception. A chair on stage is not only something convenient to sit or stand on, it is also able to tell us as audience things about the 'room' in which we see it, the time and place that the performance asks us to imagine, and things about the people who own it, and quite possibly a lot more. Similarly, a dress is not merely a covering for a body – it can also tell us about the status and habits of the character represented by the actor wearing it. These objects are open to being read by an audience – in the same way that every object on stage is. What Fischer-Lichte reminds us of here is that light too has potential for being read by an audience, for creating symbolic meaning; sometimes in ways that are deeply rooted in our cultural history.

The potential to carry symbolic meaning has increased steadily as developing technology from the late nineteenth century onwards has enabled more control of the qualities of light on stage at any particular moment. Once it became more possible to control it, many theatre directors and designers were quick to see the potential of light on stage. There is plenty of evidence that light has been an important and well-considered part of staged performance since long before the invention of the electric lightbulb.[9] However, to quote Fischer-Lichte again: 'it was not until ... the transition to more differentiated lighting had been accomplished ... that it became possible to assess and interpret the light on stage as a theatrical sign.' (ibid)

Many contemporary playwrights, directors and others want light to do more than simply illuminate the players and the scene:

> **Peter Mumford:** I think directors expect lighting designers to be artists now, and are disappointed when they're not. For the most part, the newer wave of directors certainly want that. I've had directors complain that all [their last] lighting designer did was ask, 'what do you want?'
>
> Theatre is a chemistry process in some sense. It's about bouncing ideas off each other. [As a lighting designer] you should be going, 'this is how I see it' and then you [and the director, and other designers] can disagree about it, and that's fine – you can have a conversation.
>
> Unless it [lighting design for theatre] is actually a creative art form you can't have that conversation – you're just engineering.

I'll be expanding on the case for theatre lighting design to be considered an art form later in the book. Clearly that case has to say more than simply 'because light on stage can have a symbolic function, it must be art.' Just because light is able to carry significance, designing that light is not necessarily an art. Semiotics teaches that we interpret the whole world in terms of signs and symbols, and the fictional detective Sherlock Holmes told us that almost anything can carry significance – in the right circumstances. So we are going to have to go quite a lot further if that case is to be convincingly made.

The stage is a place where we, theatre makers and audience, are engaged in a game of let's pretend. The games we play range from the child-like diversion of pure spectacle through to productions loaded with deep psychological and metaphysical significance, and much in between. For the most part, audiences buy into this willing suspension of disbelief which allows a few sticks to stand for a house, a costume to transform a well-known actor from television into a long-dead king, or a square of light on stage into a room with walls. Whatever the game on stage, as Austrian novelist and playwright Peter Handke has it, 'in the theatre, light is brightness pretending to be other brightness, a chair is a chair pretending to be another chair' (cited in *Great Reckonings in Little Rooms* (States, 1985) and frequently quoted elsewhere).

I think we need to take this idea a little bit further, however. Even in production situations when 'the brightness on stage is pretending to be' the natural light of the everyday world, as we heard from Nick Richings earlier in this chapter, 'It would be a very dull play if you just had grey daylight streaming in all the time, because it's not dramatic.' In just the same way, in most circumstances the stumbling, half-structured sentences of everyday speech would not be good enough for the stage. As the British film actor and theatre maker Steven Berkoff said in a 1998 radio interview, 'real life is theatre watered down. Life is second class theatre'[10] and by implication theatre must have more power, more intensity and more poetry than real life – and so must its light. Following playwrights, what lighting designers seek is a dramatic version of naturalism. The light is pretending to be other light, but it is also capable of doing a lot more. Using the analogy of film-making, Neil Austin give a clear account of what a lighting designer might be able to achieve, or might be expected to achieve, beyond making light that pretends to be other light:

> **Neil Austin:** If we were to translate this into the film world, the LD [fulfils the function of] many different people. Most obviously the DoP[11] or the lighting cameraman. That is the person who is choosing where the lights go and what they are pointing at. But the LD in theatre is also the DoP from the point of view of choosing the shot – is this a wide shot, a mid-shot, a close-up? Is this a super super close-up?
>
> You are also the colourist – [making choices about] the hue of an image – that has a massive massive effect. You see it constantly in television programmes now – they are making use of colourists. And you're the focus puller – in the way you can take the attention of the audience from here to there.
>
> And on top of all that you're the editor too. Because the lighting designer chooses how you get from here to there – where you're looking now and where you're looking next, and also how you get from this moment to that moment.

Neil Austin is not the only LD to have used this analogy with the film world. Peter Mumford also used the analogy, but is keen to point out its limitations:

> **Peter Mumford:** It is different to the way you do it on stage because that's always a wide shot. You're pulling people's eyes on stage by what you do with light. It's not actually changing the frame as it is with a tracking shot or a zoom or an edit in film.

All of this it is possible to do through the thoughtful and planned use of light on stage, and yet this is rarely acknowledged in writing analysing performance.

Making another good point about the importance of the role of LD as well as the role of light on stage, Austin went on to talk about the way in which he, as lighting designer, frequently coordinates all the visual and sonic elements through a scene change:

> **Neil Austin:** I find as lighting designer that quite often you are heavily involved in [what happens when in a scene change]. I never want to say directing because that sounds like you are taking responsibility away from someone else, but let's for the moment use the word. You direct scene changes, you often choose exactly how it happens – when the music comes in. Music and lights often go together. They need to motivate the changes together. One without the other sounds very strange – but also when the fly cue should happen, when this piece of scenery should move ... And also what is the focus during all of this? What are the audience supposed to be looking at during this scene change?
>
> You know, you can say to the director, 'if we just leave that person on stage a little bit longer, we can get most of the scene change done, and now we fade out [and now we're] into the next thing.' As an audience member, what you have been watching is the previous scene closing down onto this one person, then all of a sudden you're straight into the next scene.
>
> [As the LD] you've made most of those decisions, so actually you are the most central person in the production team [at these points]. And because you are on cans you are the most central because you are the one who understands what the other departments are doing, and how to solve those problems. With lots of directors I work with this is a lovely moment of collaboration – when they hand that responsibility across.

Many LDs would recognise the enhanced role Austin is talking about here, and have experience of similar situations themselves. Few other LDs would put it in quite such unrestrained terms perhaps, but then one of the issues that comes out of the transcripts of these interviews is the tendency to diffidence exhibited by most of the interviewees most of the time. For example, almost all of them talk about their own practice in the second person, frequently substituting 'you' for 'I.'

If light on stage is so important, how come it is so rarely written about?

This is a question I have asked myself, often. Here are four interlinked reasons I've come up with.

1. Writers are intimidated by the apparent technical complexity of the *machine* that is needed to get the light onto stage.
2. It is often not clear to writers what affects are due to lighting in the multichannel environment of a live performance.
3. The limitations of language – 'whereof one cannot speak, whereof one must be silent', to borrow what is probably one of the most commonly used quotations from the philosopher Wittgenstein.
4. We can't see light – only its effect on the bodies and objects that it hits.

Reason one: Writers are intimidated by the technical complexity …

This is hinted at in the earlier 1939 quote from McCandless. Theatre critics and academics seem to find 'the specialist knowledge' required to create lighting design (on all but the smallest scale) intimidating. Stage lighting often seems to the outsider to be something close to a technological dark art. The technical language used by theatre lighting specialists can seem impenetrable, and what most people see of how a lighting design is realised does little to provide an understanding of what anyone involved is actually doing. Of course this hiding of the way it works is part of what many refer to as 'the magic of theatre', and it applies to other areas of theatre practice, including acting – 'how does he or she sustain that performance every night?' for example. But with lighting there is the additional mystery of the technology as well. To paraphrase the sci-fi writer Arthur C. Clark, writing in the 1940s: Any sufficiently advanced technology is indistinguishable from magic, and the technological complexity of the systems supporting the lighting designs on some of the productions talked about later in this book is advanced. It is a big jump from most people's everyday experience of playing with light in their home or at work to contemplating how to play with the many dozens of moving lights and other fixtures that might be used to light an opera, ballet, musical or drama on a large stage today.

Compare this to fine art painting – a subject that many who can't do it are happy to write about. Most of us play with paints and crayons as children. Whether or not we then go on to study painting, drawing or any other fine art, the personal experience of 'playing' at painting gives us some understanding of how 'proper paintings' are made. The skill of many fine art painters, displayed in their rendering of aspects of the world on a canvas, or in the evocation of emotion with more abstract work, can be understood in terms of a development from the playful mark-making of childhood. In other words, even if the extent of development required to go from my childhood daubs to something like a Rembrandt self-portrait is massive, having done the childhood daubs helps me understand something of how the Rembrandt was made, and helps to give me the confidence to comment on it.

Many children play with light, showing a fascination with shadows or mirrors or the way tinted sunglasses or transparent sweet wrappers change the colour of light. However, this light play is rarely honoured by the adult world in the ways that children's painting is. It is rarely brought into the classroom, for example, and can't be stuck on the wall alongside the latest drawing. Even when a child's fascination with light is encouraged by the adult world, it is rarely given the context that would allow the child (or subsequently the grown-up) to use the experience to help them understand how stage lighting works. Too often the implicit or explicit message given is that while playing with light as a child was simple – a desk-lamp, torch or raw sunlight for example – a theatre (or concert) lighting system is complex, intimidating and dangerous, and an outsider cannot understand it.

Although almost all sighted people have an early fascination with light, something performance lighting design depends on, this fascination rarely translates into an understanding of how it can be controlled. This contributes to a lack of common understanding of the subject, which in turn inhibits talking or writing about light on stage. This in turn has contributed to a sense of intimidation among general critics and other theatre academics when it comes to writing about lighting. For those of us who do understand the 'dark art' as technology rather than magic, the way through this impasse has appeared to many to be to 'go technical', and write for a specialist rather than a general audience. Often, too, we have written about the lighting system – the instruments and technology – rather than the light itself on stage.

At this point I should make a confession: Fischer-Lichte, like Patrice Pavis in *Analysing Performance*,[12] demonstrates to me (an experienced specialist practitioner) an incomplete understanding of the practice and the technology of stage lighting design. Even after I had made efforts to become an academic, on my first reading of both these authors I discounted their useful insight because of their incomplete understanding of my specialist field of practice. This is probably a common reaction amongst specialists – especially those who like me are largely self-taught. However, it is something practitioners who wish to engage in academic discourse have to overcome. For practitioners who want to understand more about how their work affects audiences (and performers for that matter) it is also useful to engage with the thoughts of distinguished thinkers and writers in their general field – even if their understanding of our practice is imperfect.

Reason two: It's not clear to writers what affects in the audience are due to lighting ...

This relates to the difficulty of analysing the ways in which theatre lighting affects an audience, when that audience is absorbed in the multi-layered performance environment. How can one be expected to study affect attributed to the lighting alone when the combined effort of the whole production team is to make the way light is working on us as audience a seamless part of a greater whole? The attempt is a bit like trying to pick out the exact voicing of the bass line in the grand finale of a major classical symphony you are hearing for the first time, without a score, and without being able to see the orchestra. Even if you have learnt to pay attention to particular details, it is hard to do that and still allow one's self to be fully affected by the whole piece.

Much of the time theatre lighting design exists as one of many layers of potential symbolic meaning, without consciously drawing attention to itself. The planning and effort that has gone into a long lighting transition; say from clear, bright, optimistic daytime to clouded, fatalistic dusk, can be done without the audience consciously noticing any of the literally hundreds of small changes to the quality of light on stage required to make the transition. The audience may even feel colder at the end than at the beginning, and they know that a change has happened on stage, though many will be unable to describe exactly what that change is, and very few will be able to say how much that change in atmosphere was accomplished by lighting design. It is important to note that no one channel on stage works entirely in isolation. In this case, no doubt the script and the performers would help to achieve the change in atmosphere, and the transition might well be made complete by the sound designer. That said, there are times when even experienced theatre critics have been known to ascribe well-handled changes in atmosphere, achieved almost entirely through light, to the playwright, set designer or director rather than the lighting designer.

As an audience watching the drama on stage – only subconsciously aware of the slowly changing quality of light that enables us not only to see but also to better understand the story – we are generally not consciously aware of the lighting designer's work. Keeping the dynamic changes in the quality of light on stage unnoticed by the audience is not the only way of working. However, it can be, and for many lighting professionals often is, one of the most impressive and least written about of their achievements.

Reason three: The limitations of language ...

This is of course linked in many ways to reason one. Much has been made of the inadequacy of language to describe our transitory experiences in the theatre, and in our everyday lives for that matter, but a particular difficulty arises when we try to describe the ways in which light affects our experience, particularly because our brains work to normalise each new lighting environment.[13]

Even when a writer gains an understanding of how light works on stage, it can be very difficult to find words that adequately describe stage lighting and the way it affects an audience. In 2014, Kelli Zezulka, the editor of *Focus*,[14] produced a Wordle™ assembled from articles on lighting design published in the magazine and written by professional drama, dance and opera critics.[15] After *light, colour* and *lighting* (the most frequently used words) the next eight most frequently used words are: *vivid, punctuation, transition, drenching, dramatic effect, defining, sound* and *perfect.*

'Vivid', 'drenching' and 'defining' might reasonably be expected to appear in writing about light in fine art painting. 'Punctuation' and 'transition' infer the importance of dynamic change. The prominence of the word 'sound' in articles about light on stage is interesting. We will hear from several of the LDs interviewed later in the book about how they frequently work very closely with sound designers and composers. There is also a generally accepted understanding that in many instances the creation of a particular atmosphere on stage is best done when sound and light work closely together. 'Dramatic effect' probably speaks to an extension of this principle beyond lighting (and sound) design – multiple elements working together to create something that is greater than the sum of its parts perhaps?

The word 'perfect' frequently implies definitive, in much the same way that the word 'right' in 'the right light' does, and we need to question this. What place, then, does the word 'perfect' have in any writing about theatre? And it is not just theatre critics that succumb to the temptation to refer to a particularly fine production or moment of performance as perfect – practitioners do it too. In this usage, perfection usually describes a production or a moment when all the many elements of on stage come together to make something that is perhaps greater than the sum of its parts. Following Patrice Pavis and Bert O. States, using aspects of semiotics and phenomenology, we might say that every channel of meaning-making is working together to produce a somehow greater meaning.

Perhaps we can't envisage or imagine a better way of presenting this theatre moment. How else is a writer to describe that a kind of transcendental absorption in a particular moment of theatre that can remain vivid and defining for an audience long after the piece has ended (especially when newspaper critics are limited to so few words)? 'Perfect' is shorthand.

Whilst it is extremely unlikely that such 'perfect' moments can be created by lighting design alone, in my experience (and I'll admit here that I am biased) such moments are rarely possible without light that has a high degree of *rightness.* In this context, and with the understanding that what has been achieved is only one of many possible perfections, it seems 'perfect' will suffice for critics, just as 'right' will for lighting designers.

Reason four: We can't see light

This is to do with the nature of light itself and the difficulties encountered in recording it, for archiving and even for public exhibition.

The dramatic text usually remains available for analysis for future generations while the performance itself remains only as fragments in the memories of those who witnessed it. Sets, costumes and props remain for at least as long as the production is running – sometimes longer – and the artefacts are frequently archived and sometimes exhibited. Lighting, however, remains only as a trace in photographs and its archaic notation, with its potential for re-creation largely defined by the exact nature of a theatre space and its technical lighting system.

Because of the largely context-dependent nature of our perceptions of light, analysing stage lighting out of context is of limited value. Because cameras – even the best of them – don't work in the same way as human vision, camera-based recordings of performance lighting are of limited use for analysis. The detailed post-performance analysis that can be made of the text and the sets and costumes is rarely possible for light.

Just as the qualities of the performance of individual actors and dancers can be captured on camera only in part, so too with the qualities of light. However, in most attempts to capture live performance – on film or video – the focus is (quite understandably) on getting as close as possible to the experience of seeing the performer live, rather than capturing the authenticity of the lighting design. Trying to analyse a performance lighting design by watching a film or video of the live performance is – I know from experience – frustrating and ultimately of limited value.

So one of the main reasons that light on stage has had little consideration from writers and critics is that it is just too hard to see it.

A short conclusion

While those who cannot paint are happy to study painting and then to write about it, those who have no technical background in stage lighting find it hard to study it for the purpose of analysis, and harder still to write about. Those of us who do have a technical understanding have tended to write mostly for aspiring or established practitioners rather than for a general audience, and to focus on what might be called the 'craft' of lighting design, rather than the 'art' (even when the books have had *art* in the title). What follows in this book is one of several more recent attempts to make the study of light in performance, and the practice of theatre lighting design, more accessible, and to overcome these four potential barriers.

In chapter 2 my focus is on starting points for a lighting design. What do lighting designers do in the earliest stage of a production? When and how do they become involved? What are the first steps in deciding what qualities of light are likely to be right for each production? What do lighting designers learn from early set models and drawings, and from watching rehearsals?

In chapter 3 I focus on what happens in the technical rehearsal – a period of often intense activity where all the elements of a production come together to make a show. This is the time in the production process when the lighting designer is often feeling at their most exposed – 'like standing naked on a table and asking "what do you think?"', as Mark Jonathan puts it.

Chapter 4 looks in more detail at collaborations (creative and technical) creativity and art. The extracts from interviews here illustrate a range of takes on all three subjects.

Chapter 5 looks at the part that story-telling plays in the contemporary practice of lighting design for narrative theatre, and the possibilities or perhaps even expectation that the lighting designer will perform a dramaturgical role – as the guardian of the story-telling. It includes thoughts from lighting designers on their work as poetry, and on their obligation to help to entertain the audience too.

The focus of chapter 6 is lighting for dance, primarily non-narrative dance. The esteemed American lighting designer and educator Jean Rosenthal wrote that 'dancers live in light as fish live in water' (Wertenbaker, 1972), and certainly the key practitioners in this chapter, Michael Hulls, Lucy Carter and Peter Mumford, believe light is more free to be the thing itself on the dance stage than elsewhere.

The final chapter looks at time – time to make the work, and time to become able to make the work – and attempts to draw some general conclusions on 'rightness.'

This makes the book sound far more structured than it actually is. Truth is that the lighting designers don't compartmentalise these things. Their practice requires them to use a lot of different skills, knowledge and *instinct* at the same time, and this is reflected in the interviews. As a taster of how we are going to get to that final chapter, here is Paule Constable on poetry, and a lot more:

> **Paule Constable:** I think it's interesting the whole thing about poetry. With the best poetry there's a brilliant lack of fear about what is not being said. This is slightly a tangent, but I think lighting without poetry can be very pedestrian. I think that lighting is like everything in a space with an audience, everything is potentially alive in that conversation. And in poetry more than prose, every word is alive with possibility. That's where I draw an interesting analogy between the two.
>
> The brilliant thing about Cormac McCarthy as a writer is that he writes prose which is as densely beautiful as the best poetry. *The Road* is [250 pages] long – and it says so much. I think the problem with prose, and why there are so many bad books out there, is that words are cheap. And light can be cheap, but it never ever should be. It never ever should be. (laugh) It's always valuable, it's always important. So the kind of foregrounding thing about it is the way that we've all taken responsibility for telling that story. And every decision you make is important. It is the poetry. There's something that I know here instinctively. There is a very, very clear link there, about how I feel about poetry and about light.

2 Instinct as Inspiration

It is interesting to note that in the UK a concern that stage lighting should be thought about right from the beginning of the production process appeared way before the role of lighting designer was established, and even before the term 'lighting design' came into regular use. Harold Ridge (1890–1957), whose words appear in chapter 1, and Terrance Grey (1895–1987) co-founded the Festival Theatre in Cambridge in the mid-1920s, and remodelled it to create the country's first open stage.[1] Their rethinking of the stage lighting system in particular allowed them to experiment with and put into practice ways of using light on stage that followed the writings of Craig and Appia.

However, once it became usual to hire a lighting designer, it seems it also became usual to introduce that person to the creative process only at a point when most of the creative decisions had already been made. As Palmer writes (summarising Rebellato, 1999), 'although the professionalism of the role was acknowledged in the post-war period, this was at the expense of creative autonomy within the creative team' (Palmer, 2013, p. 248). For most of the lighting designers interviewed for this book, it is important to challenge this approach whenever possible, and to reintegrate the creative use of light into the earliest creative discussions. Having said that, for many this remains an aspiration much of the time, as the more usual process is for the director and set designer to begin discussions before the lighting designer joins.

To begin this discussion, here is Mark Henderson. Of all the lighting designers interviewed here, Mark has perhaps been the most successful for longest. He has certainly won more awards than any other British lighting designer: five Oliviers to date, the first in 1992 (for *Murmuring Judges* and *Long Day's Journey into Night*) and the most recent in 2010 (for *Burnt by the Sun*). He has also won the Tony for Best Lighting Design in 2006 (for *History Boys*) and has many other awards and nominations to his name:

Mark Henderson on starting points

> **Moran:** So, when you sit down to light a show, what are your starting points?

> **Mark Henderson:** I guess I'm very much driven by what I'm given as a set. That's pretty much where it all comes from. Because I normally come into a show when the director and [set] designer have already had some conversations. They've had some ideas and a scheme. I normally come in at a later point. There's the odd occasion when I'm in from the very beginning, but not very often. So you're given pretty much a fait accompli as to the space. If they ask me if I think a ceiling is a good idea, I'll usually say: 'not really' (laugh) but generally I come in a stage later than the director and designer. So whatever preconceived thing you might have in your head from reading the script, that's blown apart because you've got what you're given.

Worth noting here is Henderson's assumption that he will have read and probably re-read the script before his first meeting with the director and designer.

> **Moran:** So then is your starting point where the light would be coming from if that was a 'real' room? Are you looking for windows?

> **Mark Henderson:** Yes. You're looking for windows, that kind of thing – something to bite into. So yes, something like a window that's going to give you [motivation for] a *key*.

> **Moran:** So do you go to a lot of rehearsals?

> **Mark Henderson:** No I don't. I tend to only go right at the very end – I might only see one run-through. But prior to that I'll have had a discussion with director and designer. And a lot has to come out of that discussion. You know, if you're doing something like a big musical in the West End, with a two-month fit up, you are going to have to come up with a rig plan long before they've even started rehearsing.

> **Moran:** And that must be true in opera too?

> **Mark Henderson:** Very much so, although because you are working within an opera house repertoire system, you haven't got to specify the whole rig. You're only adding specials and things that are specific to our show – you still have to work within their system.

> **Moran:** OK. So you've had the conversations with director and designer at the beginning of rehearsals, and then you go away while they remain immersed in the piece. When you come back, how do you go about re-immersing yourself into the team – so that you can make your contribution?

> **Mark Henderson:** I'm not sure... It varies actually. With a director and designer I've worked with many times, you know Jonathan (Kent) and I have done maybe 50 or 60 shows, so we've got an understanding, and so...
>
> **Moran:** Is there much conversation?
>
> **Mark Henderson:** Yes, there's usually an amount of conversation, and Jonathan (Kent) is very visual – he knows what he wants to see from me – and what he doesn't want! Which is not the case with a lot of people. Some people don't have a visual capacity – don't take as much interest as other people. So it's very much dependent on the particular production.

Later in this interview, in a part of the conversation that was focused on what lighting does for a production, Henderson makes a substantial claim for what he gets from his brief visits to the rehearsal room:

> **Moran:** It seems to me that to do that [control of where the audience should be looking] you have to be very much inside the show. You have to have a keen understanding of who are the important people on stage at any given moment...
>
> **Henderson:** Yes – that's about focus isn't it? [directing the attention of the audience]
>
> **Moran:** Yes. How do you think you get that when all you've done is to read the play a couple of times and seen maybe one run-through? Are you channelling the director in some way?
>
> **Henderson:** No. I think it's possibly you just get that from the rehearsal room. You just get it from a quick visit. You know you should be looking there – you know you should be looking here. I don't like to spend too much time in the rehearsal room because you start nit-picking and seeing things that you shouldn't be interested in. I think in my experience, the advantage of seeing fewer [rehearsals] is that you are drawn to what you should be drawn to.

Although Henderson does not explicitly mention instinct here, I think it is clear that, like other lighting designers interviewed, he feels he needs to have an instinctive understanding of where the visual focus of the audience should be at any given moment of the piece. It is this understanding that will then inform decisions about the physical distribution of light on stage through the piece, guiding the visual focus of the audience. Henderson – like an audience member who sees the piece for the first (and usually only) time – very often relies on a single viewing of the piece to inform this understanding. The experience of having made hundreds of productions surely contributes to his understanding, as does having read and re-read the script, and taken part in creative conversations with director and designer. As we will see, however, almost all the other lighting designers interviewed strive to see much more of the rehearsal process.

Here is Jo Town, who does most of her work in drama, first talking in general terms about the beginnings of her process, and then more specifically about *Our Lady of Sligo:*

Johanna Town on starting points

Moran: So let's begin by talking about what it is that you are trying to do when you start off lighting a show.

Johanna Town: Well several things: The first thing is to be true to the play, and the director's vision for the play, which may actually be different to what I feel. I start off with what I have read and try to imagine taking the audience on a journey, whether that's a true real journey or we take them into a different imaginary world; it all depends on what the script calls for. My other starting point is how do I light the actor. These are both equally important to me, they're not in any order. ...

Moran: The way that you respond to the space, that's the designed space or the creative space?

Johanna Town: I think the brief from the director is equally the brief from the designer. I think we [lighting designers] are partly the third step. Especially in the sort of work I do which is drama, you are working to help create their vision and input into that.

Moran: I think I first came across your work as a lighting designer at the Cottesloe. *Our Lady of Sligo,*[2] set in Ireland with a single interior set. It just looked beautiful.

Johanna Town: [Making] all those demands doesn't mean it doesn't have to look beautiful. My intention on that show was to be true to the room and environment the show was set in, and equalling that with the task of keeping Sinéad [Cusack], who sat in bed 90% of the time, the main focus; that was my initial vision.

Then you go okay the play is set in Ireland, so there's an amazing quality of light about Ireland. I'd been given a lovely big window if I remember rightly at the back, it allowed me to throw shafts of light through and get the feel of the soft clear Irish light into the room.

I have an actress in a bed who has got to be constantly cheated [unnoticeably highlighted] with light in order to make sure they remain lit as well as giving the impression of the room going through night and day and that it is a real place, but most importantly also enough light to keep her in focus.

So those are all the elements you put in which basically adds up to a design. She had to remain well lit, especially through the long dialogue, otherwise you wouldn't have enjoyed that show. You would have been completely lost because the play is all words.

I do a lot of words in most of the work I do. That's not saying that somebody has to be lit the whole time because there are times when it's absolutely appropriate for somebody to be speaking in the dark.

It's about creating a reality. A lot of people would say that I don't work in reality because I'm not true to the real colour of a space. I use a lot of heightened colour and that's because I see everything in colour. I read a script and I see it in colour, I write in margins what colour emotionally I feel with that word or that scene and I do this as I read the script before I even meet the director.

Two things to note here. The first is that Town's initial description of her process is similar to Henderson, in that she reads the script before her first meeting and responds to the initial ideas of the director and designer. When she starts to talk about a specific production, however, she begins to sound much more like an active practitioner, and when she starts to talk about colour, she is moving quite a long way from traditional practice at least as it is described in most text books, which tend to suggest that the set designer defines the colour palette. The interview continued:

Moran: Do you think that the language of colour is a thing which is something you've always had available to you?

Johanna Town: It has always been there, strong colour.

Moran: Particular emotional responses to particular colours, does that change from show to show or do you think something like – if it's sad it's always blue?

> **Johanna Town:** No, it does change. It has strengths and weaknesses depending on how I'm reading the script. Sometimes I am influenced, if I've met the director before I've read a script or seen a set design before I've read a script. That will often pre-influence my choices of colour.
>
> I suppose some of my colour choices are the same though. I played with emotions through colour on a play I've just finished, *Tonight at 8:30.*[3] Light streamed through the window and in the very sad pieces I tried using green. I tried for one of the shows a smog green because it was set in the 50s, and the character talks about the smog so I had that grey-green-y smoggy sort of feel to it coming through the window effecting the emotional feel to the room.
>
> Then in another of the pieces, *Still Life,* when it's summer but the couple are breaking up and not going to see each other again, the daylight had a different tinge of green in it. It wasn't yellow, it wasn't L202 or L203, it was that yellow that's like a sort of L730 colour. The light through the windows was achieved with LED;[4] this is why I can't name the exact colour as I made it up through the LED colour mixing.
>
> I only had room for three lights through the window which then had to work on all nine plays so using LED lights was the solution.
>
> Green at the moment expresses pain for me. It probably always has, but of course some people would read that and go, 'We're not going to light that scene green?' but it's about how you push the natural real colour and hint it or shift it green, and that's what I find really interesting.

So alongside the starting points of traditional lighting design practice, key to which is serving the already agreed concept of the director and set designer, Town brings an instinctive emotional response to the text expressed, in part, through her colour choices.

Town documents her response as a cue sheet with details of what she wants the lighting to do for each scene, and this helps her decide where the equipment in her rig needs to go:

> **Johanna Town:** I very rarely submit a plan to the theatre or production electrician until my deadline so it can keep moving. Usually I've had to do the costings and placed my hire order and all of that sort of stuff, but I don't actually say it's finished until the last possible minute.
>
> I draw a plan from my cue sheets. I read the script, I write the colours in the script, I go and have meetings, and then I read the script again and

> I write up a full cue sheet. This is then edited throughout the rehearsals and then I'll draw the plan by reading the cue sheet. 90% of the time that's the cue sheet I take into tech with me.
>
> **Moran:** Seeing what you need to create each cue?
>
> **Johanna Town:** Yes, and it allows me to go on a journey through the show. You've got that emotion or action happening there, then you've got another thing happening here, somehow you've got to get from there to here, and by writing it all down on a spreadsheet I can see the journey of the lighting through the show. I suppose it's my way of doing a story-board. I can't draw, so I do my story-board in words I suppose.

This way of developing a response through the cue sheet is quite common, as we will see.

Natasha Chivers on starting points

Natasha Chivers aims to be involved in discussions earlier in the process than either Henderson or Town, who generally don't see the first, white card model of the set:

> **Moran:** Where do you aim to start in the production process? Before the final model showing, at white card or...
>
> **Natasha Chivers:** More and more at white card, though I would usually have had chats – what it's about and so on – so that when I go into the white card meeting I've got a grip of what we are trying to do, but being at the white card [meeting] is brilliant. Then it's looking at the designer's research material, learning why the choices have been made, and finding out what has been discarded. I think that's quite useful, 'well it was going to be like this but it's not and this is why,' which helps me understand how they [usually the director and set designer] have got to where they have got to.

The director and designer still come first here, but Chivers clearly strives to be on board creatively as early as possible – ensuring that she has spoken to both the director and designers before the first presentation of concrete spatial ideas (and everything that goes with these) at the white card meeting. The idea of gaining an understanding of the vision of the director and designers in part through knowing what they have rejected is interesting here and illustrates just how keen Chivers is to influence the joint understanding of the piece with the director and designer. The interview continued:

> **Natasha Chivers:** But I do think I'm hugely instinctive really.
>
> **Moran:** Any handle on where that instinct comes from?
>
> **Natasha Chivers:** Not really. I think I've always been very curious – I think I'm quite a sensitive person. Sometimes it can take me a long time to come up with something – but that doesn't really hinder me. I think instinct for me is just that.
>
> **Moran:** What part does any attempt at Naturalism play in your early concepts? Are you driven by, for example, making the light appear to be coming through the window, and be the right colour for the time of year?
>
> **Natasha Chivers:** I think it is often a starting point, even if it is then discarded. I've had meetings with directors when they have been a bit worried, for example when we've already agreed not to do it in a natural style, and I ask: 'what time of day is it?' and they think we're about to do some kind of really awful [imitation thing]. But I think it's about knowing those things like what time of day it is, and then making a decision about whether they are relevant or not. Or what to do with these facts. Because you have to start from somewhere don't you? I think I'm very driven by the architecture of the space. I try to get a really strong sense of the space, and you might end up feeling that you want to push light in from one side. I've done a lot of site-specific work and that may be where a lot of that kind of idea comes from. I think that's probably part of that initial dialogue (with director and set designer) and I think that's probably more of a starting point, than anything else.

This idea of the architecture of the space driving choices is a notion that Adolphe Appia and Gordon Craig both thought important. Later in this chapter we will hear from Bruno Poet and Jon Clark on this subject. Returning to Chivers, however, the next passage throws some light on how her earlier practice making site-specific work influences her process when making more traditional theatre:

> **Moran:** It seems to me in your work that you are often very keen on the effect of light on surfaces – perhaps going back to the eureka moment in the Almeida with Simon Corder.[5] Is that something you often plan for from an early stage or is it an accident – you see the possibilities of it and decide to 'go' with it?

Natasha Chivers: Is it an accident or design? ... I think, when I do a site-specific project – or when I first see a model box – all sorts of things are happening that if you sit down afterwards you might be able to unpick those thoughts. But in the moment it's an instinctive response to some key features. It might be to a skylight or a particular wall.

Moran: In that moment do you find yourself explaining to others what those key features are?

Natasha Chivers: I find I communicate less and less actually, but not in a way that's a problem. I mean, that's one of the great things about being a bit older and having an established career; you get given a lot more freedom, and I tend to choose to work with people who will give me that freedom and space, because I find it more creative.

So with a new person, I might give an outline of what I think I'll do with a piece.

[With site-specific work] half of what you're doing is lighting the space, and half of it is lighting the piece. In a normal play, nearly all of what you're doing is lighting the piece. For a site, the space becomes part of the narrative of the piece, so I guess you're still lighting the piece when you're lighting the space.

This feels to me like an important point – narrative can come from the space as much as from the spoken text or anything else, and light can help space tell its stories. I follow Appia in arguing that this is true for some constructed spaces (sets) as much as it is for the spaces of site-specific work:

Natasha Chivers: Jon Clark and I used to do a lot of this kind of work together... We would just hire a lot of equipment that we knew would be suitable – little manual desks and dimmer packs, birdies and sun-floods [small lightweight kit that can easily be hidden or screwed to the wall or floor of the site-specific venue]

[We got good at working in a way where we accepted that] we're not going to know what we're doing till much before [the show opened] because they were not going to have blocked it before then, so we had to start, going round saying, 'you do that room and I'll do this room,' and then talk through what we are going to do about this wall, or that space – and that's just heaven isn't it? and fun and random.

And then once you've done one of those, you've had enough because there's no infrastructure, etc. So you go back to a theatre. Then when

> you've forgotten what a pain site-specific work is, and you begin to get a little bored of theatres again, you go back and do site work again.

Chivers calls her response instinctive, but it is important to note just how much effort she goes to in order to get into the same creative mind-set as the director and other designers. This then gives her the freedom to let her well-informed instincts - well informed about the creative and technical aspects of her practice but also well informed about the creative spaces of this production - drive her work. This is a good description of a possible starting point for an active practitioner.

Paule Constable on starting points

Paule Constable is perhaps the best-known lighting designer in the UK at the moment, with a distinctive style and way of working. Part of that style is to do with having the right unit in the right place, something that can only be achieved consistently through close collaboration with all the departments from very early in the process:

> **Moran:** For a lot of your work that I've seen, particularly when it's been large scale, it seems to me that it is often about getting one or two key units in exactly the right place.
>
> **Paule Constable:** It is - emphatically it is. Or the right dynamic. In something like *War Horse*,[6] it's about a lot of units in the right place - it's about a big push. Yes. I'd rather have two good ideas and take them to an absolutely extreme place, than have 50 half ideas.
>
> **Moran:** For *Danton's Death*,[7] with the light coming through that huge window?
>
> **Paule Constable:** Yes. That's a good example of how far you can push an idea. In the play it rapidly became clear that there was never an interior when the windows wouldn't be covered with shutters. So we were only concerned with the light coming through the opening or seeing the windows covered up, which meant that we didn't need glass. So there was nothing in them. That was important because it was all about making the most of an aperture - so you want maximum transmission and minimum reflection. The set designer was Christopher Oram, and he said: 'should we use gauze? Should we use Plexiglas?' And I went: 'why don't we use nothing?' And for the audience, I don't think they ever questioned that but it meant that we got the most beautiful thick light coming through the windows.

Moran: That point of engagement – right at the start of the process often well before even a white card model meeting – is quite unusual, and a lot of lighting designers say they would kill to have it. How do you think you have managed to establish that?

Paule Constable: I think it's because I don't work with many [different] people. So I think that the people I do work with either get used to it, or have learned to deal with it (laugh). It's probably a mixture of both. But I think it's also – well everything in the world we work in is about casting, and that's from your lead actor to your stage manager, and to your lighting designer. You know, I've learned over the years that there are people I shouldn't work with, because they just want me to come in and turn the lights on and off. And I'm not good at that. I mean, going back to the early days of Complicite, we all just did what we did, and the fact that I responded in light ultimately was neither here nor there ... so I'm not used to being pigeon-holed – to being told that that's my corner. I'm very respectful – I hope – of what people do, but I love to be involved inside a piece; to understand what is making it tick.

There are things you learn with experience – the model you see at that first meeting, the 1:1 version of that, is what you're going to have to live with when you hit the stage. So any moment that you miss, the decisions you don't make then will haunt you.

It's great to learn from the rigours of touring and the one-dayers[8] because the mistakes you make, things you miss, you can put right next time. The things you miss when you're doing the first meeting about an opera which is in three years' time, the things you miss then you really are going to have to live with them – and it's your fault if you don't discuss them with the rest of the creative team. A director and designer may say: 'We want it to be a three-sided box, and for the light to be this, and this and this.' If you don't take that on board – if you don't say: 'well, what this means is this is how we can use light in this space' – if you don't have that conversation, or if you don't go: 'OK – if you really want us to still be able to break it up we're going to have to put in a position here, and here or here...', or 'you've got windows all the way down one side, let's look at making that a real way in for the light.' If you don't do that ... then you haven't started the conversations early enough.

It's all about trying to make my life easier and about managing expectation.

It is clear, then, that Constable sees early involvement and initial meetings and discussions as key to being able to do the work she wants to do. As with most of those interviewed here, she also likes to spend a lot of time in rehearsal:

> **Moran:** And you try to spend [a lot of] time in rehearsals?
>
> **Paule Constable:** Yes. I think one of the things about doing lots of opera is that you get to know the repertoire very well. I know *Don Giovanni* very well, and *Cosi Fan Tutte* – I've just done my fourth *Cosi* – and I'm very happy to do it. It's part of the pleasure – learning about these great pieces of art!
>
> And I really spend time in rehearsals, because I can't imagine doing a lighting session sat next to a director and having to ask: 'right, what happens now?' Because I know 'what happens now', I can just rough something in – the director can keep working, I can keep working...
>
> **Moran:** Yes, because that raises something I've talked to several LDs about – how you get to the same understanding of the show as all the people who have been with it constantly – the director, designer, stage managers and performers – when you last saw them several weeks ago?
>
> **Paule Constable:** I think you also learn that there are projects – like *Light Princess*[9] or *Curious Incident*[10] – where you know from the beginning that these are shows that you need to be around a lot. So you kind of see things coming. Someone like Michael (Grandage) will say 'I really need you for the following time'; he wants you in and out through the rehearsals, and then to watch the runs. Whereas with someone like Marianne Elliott you just spend as much time as you can, just getting it under your skin in every way possible. Rufus (Norris) as well – the longer you spend with them in rehearsals the better, because then you are all making it, together. But yes, anticipating that is hard.

Later on we will hear from Constable that her process is different when the director knows how they are going to direct the piece, and that this is not always the case. This is an extract from a conversation between Professor Simon Shepherd and Michael Grandage, published in 2012:

> My view is that one of the most important things to provide as a director is leadership and a vision of some kind. ... You draw together a collaborative team of designers, lighting designer, costume and scenic designer (if they're separate), a composer, a choreographer, a fight director, each

> where appropriate... a whole team before the actors are even on board. With them, you come up with a vision for the piece. In doing that I hope to excite and motivate the team, and particularly the designer in the first instance, to start coming up with something which informs that vision and then goes into a collaborative place that takes it on and develops it in other areas that you weren't expecting. So leading that first off is the most important thing. (Shepherd, 2012, p. 189)

In contrast, Marianne Elliot, the director of *The Light Princess* and *The Curious Incident of the Dog in the Night-Time,* has a reputation for encouraging a process that develops the piece in the rehearsal room, with whoever is there, and 'the longer the time you spend with them in rehearsals the better, because then *you are all making it, together*.' (Constable, 2014 – my emphasis)

Ben Ormerod on the rehearsal room and developing a structure

Paule Constable and Bruno Poet are just two lighting designers who cite Ben Ormerod as a significant influence on them as artists. Here Ormerod returns the compliment with an insight gained from his one-time assistant and associate Bruno Poet:

> **Ben Ormerod:** Bruno put it really well. Bruno, I know, spends a lot of time in rehearsals. He said, 'You've got to spend hours thinking about the play; where better place to do it than in the rehearsal room?' He's absolutely right. I find it difficult to concentrate, generally, unless I'm at a production desk lighting. So, when I'm in the rehearsal room where people are actually doing the play I find myself thinking about almost nothing else. That's fantastic. I try to collaborate by knowing the production so closely and what it is that everybody's trying to do, that we don't need to talk too much about it.

Earlier in the interview Ormerod talks about how his process starts with a search for structure in the text on which to build the temporal structure of his lighting design:

> **Ben Ormerod:** For me, what is at the heart of a good lighting design is a sense of structure. It's absolutely key to lighting a show, being able to understand how to structure the lighting design, and the structure of the lighting design is going to be generated by the text. That's something you can do outside the rehearsal room, because you've got the text. That isn't likely to change a great deal in the rehearsal room. Because of my musical background, and through studying composition

techniques, I've found that understanding how to structure a lighting design across time, in the way a piece of music is, gives you endless ways of generating materials with limited resources, which is a big part of what musical composition is all about.

Explicit in the notion expressed here is that for Ormerod at least, gaining an understanding of the deep structure of a piece is key to creating a responsive lighting design that supports the work on stage. In this chapter my concern is with starting points, so here is an extended passage from Ormerod on just that:

> **Moran:** Where do you start with a lighting design project?
>
> **Ben Ormerod:** Okay, it depends on the project. For instance, I'm doing *Tristan und Isolde*[11] in almost exactly a year's time and I started work on the score two or three months ago. That is because for me personally Wagner requires an enormous amount of preparation. I have to know the music inside out, particularly for Wagner, almost more than anybody else; Berg, also, I would say. Harrison Birtwistle, again, I would say. When I lit *Mask of Orpheus* I worked on the score for months before we went into rehearsals, so that I absolutely knew my way around it.
>
> So, the preparation is key and the preparation, for me, is about two things. The first thing is about understanding what the relationship with light is, so there's an obvious thing; let's not have sunlight pouring through the windows when it's supposed to be raining, for instance, stupid things like that, let's get a sense of what time of day is it, what's the weather like, those kinds of things.
>
> But there's something else; does light play a structural role in the text? For instance, in a play like *Hamlet*, it doesn't. In a play like *Macbeth*, it does.
>
> Sometimes when I read a play for the first time I get a sense that light might actually be embedded in the text somewhere. For instance, it really is in *Macbeth*. It's obvious as soon as you start reading it that light is one of the key images; in the same way that commerce is a key image in *The Taming of the Shrew*, light is a key image in *Macbeth*.
>
> So this is the second stage of the preparation; I type out all of the references to light in the play – and this is a trick I learnt from Alison Chitty,[12] who once said, 'Write a list of all the props in the play and then read that list as if it was a poem about the play.' That's what I do. I make a list of all the light references in the play and then I treat that list as a poem. The

'light poem' about *Macbeth* runs to about eight pages on A4. Sometimes there are entire speeches dealing with light.

When you read this list you find that there is this hidden story about a battle between the day and the night; it's a very clear and unwavering story, with no internal contradictions. There is this progressive infection of the day by the night that takes place in the play, which is brought about by the spells, the incantations of Macbeth and Lady Macbeth. 'Come, thick night, And pall thee in the dunnest smoke of hell, That my keen knife see not the wound it makes.' Its pivot is the Banquet scene;[13] and one interesting thing about that scene is that it ends on the hinge of the night and the day. 'What is the night?' 'Almost at odds with morning, which is which.'

Almost to a line, the role of Macbeth falls into three equally sized parts; from the Blasted Heath to the discovery of the dead king; from the first time we see Macbeth the King to the end of the banquet scene; and from the apparitions to the death of Macbeth. The middle third of the role of Macbeth is entirely concerned with the Banquo story.

It's a big chunk of the play and it contains a surprising number of references to the time of day it's going to be when Banquo leaves the castle; Shakespeare wants you to understand that he is going to leave towards the end of the day. It's light at the beginning of the scene when Banquo is murdered, but by the time he comes on, they're calling for torches. Shakespeare is so clear about this. And of course the banquet scene takes place that same night, and finishes at the other end of it. The fact that Shakespeare takes such pains to set key events on the cusp of the day and the night is a clue as to the role that light could play in the production.

The absence of those clues is, in itself, informative as well. When you take a play like *Hamlet*, you discover that there's nothing like that. *Macbeth* is so short and yet has got all this stuff about light and then *Hamlet*, which is so long, has got a page maybe, the odd reference: 'Will you go out of the air?' Is that a light reference? I suppose it is. So, there's that.

That's how I get started. Then, somewhere in there, there's the usual viewing of the model box and sometimes I'll see the model box for the first time with the actors, which is quite scary, and sometimes I'll see the model box at a very early stage, when it's being designed. That's for all sorts of reasons. Sometimes you don't get booked until after the set's been designed, amazingly enough. Sometimes, you're actually not in the country when the set's being designed.

This level of analysis of a text to reveal a structure is unusual but not unique, especially for lighting designers working on classical texts and opera. As Paule Constable said: 'I can't imagine doing a lighting session sat next to a director and having to ask: "right, what happens now?"' As we will see in the next chapter, the way contemporary lighting designers work – lighting over the technical and other stage rehearsals as opposed to having dedicated lighting sessions – relies on them having a detailed understanding of the piece and the shared vision of how the team want that piece to work for the audience. However, most of the other lighting designers interviewed for this project aim to get that understanding of the structure of the text from the rehearsal room rather than from their own detailed analysis.

Bruno Poet on starting points

We have already heard from Ormerod that Bruno Poet considers the rehearsal room to be a good place to think about and develop his creative response to a production. Here we hear Poet talking about his own starting points and how he uses his observations of rehearsals as one of several stimuli to develop his response:

> **Bruno Poet:** I think it is really hard to generalise. I think I have an instinctive response to a combination of the scenic environment, the music or the words, and the shape of the performers on the stage. And so I think a lot of the right light is about the angle. I think that's the thing I think of first, really. When you're designing a plan and watching rehearsals, I think the first thing I think of before anything else really is the angles – the shape. Keying the scene, I suppose. It's not always the key light in the traditional [text-book[14]] sense. It could be a three-quarter backlight, for instance.
>
> I tend to do pretty much everything in my head, design-wise. So I tend to just sort of look at the model, and watch rehearsals. I don't do particularly complicated drawings or notes and things. I suppose I think of the possibilities from looking at the set model, and then I juggle those possibilities when I'm watching rehearsals. And then through that, I come to some kind of conclusion about what I feel is the right shape for each scene. I guess what the shape of the scene should be dictates what the light should be.
>
> **Moran:** Are you talking about physical shape and also about the shape through time?
>
> **Bruno Poet:** Yes. I think that's it. Some things are much more literal in terms of, you actually need to represent a naturalistic time of day. Some

things aren't. But I think actually whether you are naturalistic or not ... I guess I still have an instinctive sense of what the shape should be, so what the key direction of the light should be.

I always think that scenes look best with the fewest lights on, really. So once you've got that first light right (and by light, I don't necessarily mean one single source. It could be 20 fixtures all going in the same direction) but it's that shape that gives me the structure for that moment in the show, which you then build from. And I suppose the interest in lighting is how you move from one shape to another – from cue to cue or one scene to the next.

Moran: To what extent are you starting off with where the light would be coming from, were this a 'real' place?

Bruno Poet: Looking at the set is a huge part of what I do. When I'm looking at the set [model] I'm thinking about possibilities. Where are the interesting angles? How can we sculpt the people? How can we make the set come to life? How do we balance keeping the actors vibrant and in focus with the set that's in the background – or not – and how those things work together or if they don't work together, how do I fix that in lighting terms?

So Poet is interested in the possibilities of where he can get light that will do interesting things into the space. Whether or not there is a *naturalistic* motivation for it is sometimes less important. In each scene, Poet is looking for a shape, something that will become the physical starting point for his lighting design when the production moves from the rehearsal room to the stage. That 'shape' comes from both the usually static set and the almost always moving performers observed in the rehearsal room as they develop a performance of the text – the developing architecture Chivers spoke about.

Jon Clark in the rehearsal room

There are similarities to Poet's approach in what Jon Clark has to say about starting points:

Jon Clark: I think I see directions more than anything, and I kind of always start from there. ...

[In the rehearsal room] I always story-board the blocking. I think it's a kind of security blanket. I think it makes me feel busy and look busy but it also engages my mind so it's not that 2 or 3 o'clock in the afternoon thing. I photocopy the outline of the set – and I tend to, almost like a DSM, story-board the main action, because when I look back over it, that gives me a good sense of shape of the scene or a dance piece. You can very quickly see the shape of the choreography when you're looking back at it. I find that's the second part of that day, to analyse [those rough drawings]. It starts to inform what you are doing and if you go into rehearsals soon enough, then that gives me the confidence to feel like I'm investing in it and I'm part of that process as well. It puts you in the room, so you're there.

Again, there is the idea of looking for an underlying shape that will inform future choices about the direction of the key light sources on stage for that scene. Here also is the implicit idea that the lighting designer will in some way absorb the structures that will underpin the production through being present in rehearsals, and perhaps this is at least part of what Chivers and Poet refer to as their 'instinct'.

Lucy Carter on starting points

The lighting designer's instinct must to some extent be influenced by training (or early experience for those with little or no formal training), and this perhaps goes some way to explaining why the approach of both Lucy Carter and particularly Michael Hull include some marked differences when compared to that of most of the rest of the lighting designers interviewed. Hulls works exclusively in dance, and much of chapter 6 focuses on his practice. Lucy Carter trained as a choreographer and started work on lighting for dance, though she now works in drama and opera too. Here is what she has to say about her starting points:

Moran: So you want to be in before white card.

Lucy Carter Yes. Definitely I want to be in as early as possible, from the beginning. ... It pays off massively if you're there at the beginning, even if, after that, they go into a huddle. It really, really, really helps.

Moran: Do you manage to do that in drama as well as dance?

Lucy Carter: More and more. [For example] the *Medea*[15] that we're working on now, I was in at the beginning and for the first couple of meetings and then they moved on and did some of the set design and stuff and then we came back in together. Yes, more and more. ...

Once there's a physical white card I need to ensure that I'm not going to be completely restricted as to where I can get light in, and I try not to restrict them either by saying 'well, there's absolutely no top light,' because that presents its own interesting challenges: not being able to have top light or not being able to have [some other angle].

Also looking at what they've designed and saying 'actually I can't even light that because you've not given me room' or 'that object, or that piece of set that you've designed needs to be side-lit in order for it to look amazing but I've got no side-lights possibilities'. So making sure that what they're physically going to put on stage isn't too restrictive, and that I can also service it, that I can also light it.

Then thinking about what our concepts were originally and saying 'well, I can't really service that original concept' or 'if you did it like this then that would be brilliant because then we can serve the original concept in this way'. So it's always a collaboration, it's always I would say a two-way if not a four-way or five-way conversation, and if you've begun early enough and you're not rushed to produce the [lighting] design, you can have time in between to go away and think about it and not fire-fight. The worst thing about not being brought in at a white card point even is that in the end all you can do is fire-fight and try and solve problems that were out of your control.

This echoes something that Paule Constable said: 'the decisions you don't make then [at white card] will haunt you'. What does Carter's process look like once the production moves into a rehearsal room?

Lucy Carter: People say to me 'have you got one great idea that you've always wanted to try? Maybe we could try it?' But I don't agree with that. I think it has to come from the idea of the piece and each piece is different and if a piece demands some side-light then it's really clear as soon as you get into rehearsals. If I find myself thinking about what equipment is available before I've decided how it's going to feel and look, before I have certain really visual images in my head, I have to kind of reprimand myself because I'm a big believer in it being about the idea, not being about the equipment that's available. Some pieces just demand amazing top light that feels really industrial and flat. If that's right for the piece, then that's right for the piece.

Moran: [Do you see the way you work as being heavily influenced] by your training, studying choreography?

Lucy Carter: Yes, I do. Even now I see a cue in a body, I don't hear the cue in the music. It's been commented on by choreographers that I work with, that they find that kind of unique – that I'm responding to the

> choreography and not musically. I do work very musically as well, I find music very inspiring in terms of the emotional quality of the lighting but I think in terms of cueing it, it's finding the right cue point and the right time for that cue so the dynamics of the lighting design come from the choreography.

Carter's starting points when she is looking for a structure appear quite different to those lighting designers whose background is much more deeply rooted in the traditions of realism and naturalism. The end result may not be that different – if Ormerod finds motivation for his lighting response to the banquet scene in *Macbeth* from the prose poem pages of references to light in the play, Carter may find her motivation in the bodies of the actors rehearsing the scene. Both will also bring extensive background research – their own and that of their collaborators. The structures each arrives at may not be that different from each other or they may be very different – but they will both have come from inside the process rather than being imposed upon the production from outside. Methods may be different, but for both, practice is built on active collaboration. As Carter went on to say later in her interview:

> **Lucy Carter:** What I do [in the tech] I feel is effortless and I enjoy it. I sit at my production desk and almost it's an instinct to decide what to do. Afterwards I think, but that's just my version, there are many other versions that could have been, it was my version that I created in the moment. ... The next day when you look at something hopefully 70% of what you did the day before is good and then there is the other 30% you think 'oh I could do that differently', so you do. So you know that actually in the moment, if you just work instinctively from your research, from the rig that you've designed, from the ideas that you've had, from the story you're trying to tell, from the conversations you've had with the director, if you make a choice in the moment mostly it's going to work really well, but there are other versions you could try.

There is no single definitive *right light*. There is only the light you make in that moment, informed by research, collaboration and instinct, and the eye of you as creative artist and integrative practitioner.

Peter Mumford on lighting design as 'the last creative act ...'

Everything the lighting designers have said here is about ways of immersing themselves in the production in order to make instinctive choices that are right for the space, the piece, the performers, the story, the overall concept of the creative team, whilst at the same time having the creative autonomy necessary to do work as an artist:

> **Peter Mumford:** If I'm describing lighting design, I often describe it as the last creative act in the process of making theatre – whatever it is. But the lighting design is the kind of melding of surface design with performance. You have a duty to that, to both those things, and to that material itself, the play or the ballet. You can never really separate the light through the window and what the people look like in the room, because they're in the room... The light through the window is very rarely an entity on its own, because if you've got a window on the side of the set, or on the back, then you say: 'ok my big thing is the light that comes in through the window, but we have to light the people in the room, and the light through the window also bounces off the walls, etcetera'. So you can take that, and use that to amplify a situation. Again, controlled naturalism. You take a reality and then you take it on a stage in order to illustrate what you want to illustrate in the piece.

> **Moran:** So that balance of three things: the appropriate visibility of the actor, the honouring of the architecture of the set and the honouring of the piece – getting that right defines when its right for you?

> **Peter Mumford:** Yes it does. But all of those things are open to interpretation. They are open to my interpretation in a sense, because the set designer gives me a canvas to work on, and that's not to deprecate the role of the set, but for me, that's my canvas. So if they give me the ceiling I'll work with the ceiling, if they give me a window, I'll work with a window. But they are basically giving me a surface that does not exist until I light it.

3 Tech: A Cauldron of Potential

For most productions of plays, musicals and opera, at some point the performers and the director move from the rehearsal rooms into the performance space, usually to begin the technical rehearsals or 'tech'. Between this moment and opening night the lighting designer's primary concern is to create appropriate looks and to ensure they are recorded in such a way as to make it possible to reproduce those same looks accurately at the right time in every performance. In other words, to ensure that each moment of each performance has the *right light*. For the lighting designers interviewed here this requires aesthetic and technical understanding and engagement, and a lot more. Often the paper or digital plan for the lighting system (and the consequent realisation of the system in the performance space) is referred to as 'the lighting design'. Rick Fisher, in conversation with a student lighting designer, gives a clear explanation that there is a lot more to the lighting design than getting your lighting equipment mounted where you want it; that is, getting the rig up:

> **Rick Fisher:** [When you plan your lighting rig] you are not rigging ideas, you are rigging opportunities to make sure you get a lot of flexibility in each area [of the stage]. [Making sure you have] different types of sources, different kinds of colour, different qualities of light, [so you can] choose what you want to use.

It is worth noting that this approach is not universal, as we will hear from Nick Richings later in this chapter and Paule Constable too. Here are some of Bruno Poet's thoughts on what is going on when he is creating lighting cues during a technical rehearsal:

> **Bruno Poet:** I think when I'm making cues in a tech – which is where we do most of the lighting – I think it's an instinctive reaction to what I feel is right. So you're guided by the set, you're guided by the director, guided by the play. And you are guided very much by what you see on stage in front of you. And you somehow add all those elements together, mix in your own experience, and decide what's right.

By the time the director and cast get to the stage, it is usual for the lighting designer to have already been at work there for some time, and for the production team to have been there even longer, building the set and preparing the rig for the lighting designer to work with, for example. Lighting designers working at this level do not generally get involved in physically getting the lighting instruments into position and connected to the supporting technology, though most have done these jobs earlier in their careers.

A common thread running through all the interviews cited in this book is a concern with the direction and angle of the light hitting performers and objects on stage. This cannot be achieved without a good deal of attention to the more technical side of the process, and an active engagement with many other aspects of the production. No matter how removed the lighting designer becomes from the rigging work, they will usually still concern themselves with the detail of what each lighting instrument will be and where it will be placed in relation to the performance space.

Expectations and constraints

Every piece of theatre is made within a context of expectations and constraints. Some are usually clear to everyone involved; maybe the types of audience the piece is to be marketed towards or the specifics of the performance space. Some become evident with experience. As Paule Constable mentioned earlier, the potential constraints posed by the set design that you don't recognise early enough will come back to haunt you later. Here Paul Pyant talks about other constraints that need to be considered:

> **Paul Pyant:** You always approach any project, whatever you do, with a list of questions: where are you doing it? Who are you doing it for and how do they do it? [With experience] you know how much you can push a company and what you can ask for. You start with the idea, you then try and equate that idea with where we're doing it, the budget, the time and the people you have involved, whether you're at the Coliseum, National Theatre or Birmingham Rep – you play to their strengths.

Days before this interview Paul Pyant had won the Olivier for his work on the West End musical spectacular *Charlie and the Chocolate Factory*.[1] The production had a rig with a great deal of technology, and a team of specialists involved in installing it and keeping it all functioning correctly:

> **Paul Pyant:** I now, certainly, have to have people around me that I know and trust and who know what I do

> ... These days you have to have a production electrician and a programmer ... I only choose those ones that I like and I know do the job phenomenally well. They'll cross the t's and dot the i's all before I get into the theatre. So once I'm in there and focusing, we know it's up there and running.
>
> **Moran:** So you've got a machine that totally works before you start?
>
> **Paul Pyant:** Totally and utterly.

Pyant articulates the aim of all those interviewed: the team get the lighting instruments into the right places and assemble all the necessary technical infrastructure to ensure these instruments all work correctly. Once the lighting designer arrives at the theatre, she or he has their machine, the lighting rig, up and working ready for them to 'focus'. This is the process of ensuring each lighting instrument is pointing in the correct direction and that its beam is shaped and coloured correctly too. Directing the technical team behind the lanterns is a specialist task, usually undertaken by the lighting designer. It is often called 'calling the focus' and covered in detail in many lighting design handbooks.

Only once the rig is up, working and focused is it possible to begin to make a creative response to the production with light, to create the *looks* on stage made from volumes of light in space.

What is the LD doing?

> **Natasha Chivers:** I think in general there's an expectation that lighting takes – not no time, but almost no time. I think it's consistently a shock to people, other production and creative team members, that it takes some time.

This section is a short diversion to provide some context for readers unfamiliar with the details of contemporary lighting design practice, to try and answer the question implied by Natasha Chivers, namely 'why does it take you so long?' Those readers who are already familiar with contemporary lighting design practice in the UK may wish to skip on to the section headed S Framework of Cue Points.

From a technical perspective, almost all contemporary lighting designs for plays, musicals, opera and dance are built primarily from lighting 'cue states' and the transitions between them. Each cue state is stored in a lighting console – a dedicated computer, which at the level that those interviewed work almost all the time is programmed by a skilled professional – the lighting programmer. Lighting cue states usually consist of the recorded intensity of a number of lighting fixtures, and if the fixtures are, for example, moving lights, other parameters such as position and colour too. Usually the lighting cue state is a snap-shot of the output of the lighting rig – a way of capturing and recreating later the particular lighting picture on stage at that moment, the 'look'. Sometimes a particular look is made up of several cue states, and often the look is dynamic – a transition between cue states rather than a set of static intensities and other parameters.

The moment in the piece when the cue state should be triggered is the 'cue point'. This is very often associated with a particular word or phrase in the text or music, but it may also be associated with a particular movement by an actor, or with another technical cue point such as the cue to fly in or move a piece of scenery. In Anglo-American theatre all the cue points are documented in a master script ('The Book') by the deputy stage manager (DSM) or show caller. For each performance the DSM or show caller uses The Book to tell all the technical operators (and sometimes the performers too) when to initiate their cues. The standard protocol for these instructions requires that each technical cue point for each operator has a number. Lighting cue points are usually designated with the prefix 'Lx'. So the fifth lighting cue state in a show will usually be called 'Lx Cue Five'. It will be initiated by the lighting operator when the DSM or show caller sees that the performance has reached the appropriate cue point, and gives the command: 'Lx Five GO'.

The how and why of this well-tried system is extensively covered in text books on theatre stage management and theatre lighting design. What is important to note here is that each static look has an associated lighting cue state, which in turn has an associated cue point, and that both the cue state and the cue point have a number – most often the same one.

The process of turning on and adjusting the intensity and other parameters of the lighting fixtures to create looks is known as 'plotting'. The act of recording a lighting cue state with its associated cue times in the lighting console is often referred to as 'plotting a cue'.[2]

Creating the look: technology in the service of art

At the scale that the lighting designers here work it is now very uncommon for them to programme or operate the lighting console. Current practice is for the lighting designer to sit at a production desk in the theatre auditorium during the technical rehearsals, alongside or at least close to the director and other members of the creative team. The lighting programmer may also be close by in the auditorium, or somewhere that gives the programmer a different view, or in an operating booth. (Most contemporary lighting desks are relatively easy to move from the dedicated sound-proofed operating position to other parts of the theatre.) In any case it is usual for the lighting designer and the lighting programmer to communicate using headsets – to improve the clarity of their communication and to minimise the distraction their talk might otherwise cause.

In the lighting console, each lighting 'cue state' is a set of stored numbers that when recalled is able to turn on some or all of the lighting fixtures in the rig at specified intensities (with a specified colour, beam quality and other attributes if the fixtures are, for example, moving lights). On stage each lighting look that the lighting cue state recalls consists of one or more volumes of space filled with light of a particular quality or qualities, from a particular direction or directions, and the consequent volumes of darkness that defines the edges and surroundings of the lit space or spaces. The same look will often reveal a performer in different ways depending on where and how that performer is occupying the space – their orientation and the direction they are looking, for example. (As I have said, this has not always been the case; the object of many lighting designers working in a traditional way being to ensure as far as possible that performers look the same wherever they are on stage – whether near the window with the bright sunlight streaming in or in the opposite supposedly dark corner of the room.) Similarly, in order to maintain the particular look on a performer who moves slowly through a stage space it may be necessary to plot several cue states and transitions, creating subtle changes to the lighting system that the audience will not be consciously aware of.

To create each look the lighting designer asks the programmer to turn on one or more lighting fixtures at a particular intensity, and when she or he is happy with the stage picture, to plot it as a particular lighting cue – that is, to record a lighting cue state. Whenever there are lighting fixtures that can be programmed to move, change colour or otherwise alter the quality and shape of the light beam they produce on stage, these things too must be communicated to the lighting programmer, so that all this information can also be recorded in the lighting cue state.

One of the significant challenges for any stage lighting designer is to engage at the same time with the aesthetic concerns required to create each look and with the technical considerations – particularly the many numbers that are associated with the computer-recorded lighting cue state.[3]

The mediation of the lighting programmer in this process can help the lighting designer to keep their focus on the aesthetics on stage, while the programmer handles the numbers in the computer. All of our lighting designers here acknowledge the importance of the lighting programmer in their process. This is covered in more detail in chapter 4.

Increasingly, lighting designers make use of the technology available within the latest lighting consoles to eliminate one or more of the sets of numbers they have to deal with. Nick Richings is not unusual in naming lighting cues, for example:

> **Nick Richings:** So there is a basic spread sheet, with all the cues, and a name for each cue, so I no longer have to think about numbers… numbers were hopeless for me.

A number of instruments can be grouped together for the purposes of plotting and this group can be given a name by the lighting designer, thus removing the need to recall quite as many individual channel numbers. Colours that are frequently used can be named too, as can other attributes of moving lights and other remotely programmable fixtures.

Transitions, cue times and dynamic looks

The aesthetics of the dynamic look of a transition such as might evoke a sunset or suggest the onset of a mood of depression or resignation is very often just as important as that of the static looks created by each stored lighting cue state before it or after it. It is the timing of each element in the transition – the 'cue time' – that will largely determine the aesthetic of a dynamic look.

Imagine a look that is largely red light changing to another that is largely blue light. There are several ways in which the cue times can be programmed. If the time it takes for the red light to lose intensity ('fade out') is the same as that for the blue light to rise in intensity ('fade in'), we have what is generally called a straight 'cross-fade'. I'm sure that you can imagine how different a two-second and a two-minute transition of this kind might be. Now imagine that we delay the fade in of the blue light until almost all the red light has gone, creating a very dark look in the middle of the transition. Or imagine a transition where we arrange to turn the stage from red to blue beginning in one corner and then spreading the change across the whole stage. Again, any of these transitions will create quite different dynamic looks depending the time taken for each fade in or out, any delays and the overall time taken. For many lighting designers it is often fine adjustments to cue times that do more to make the light right than adjustments to intensity. Smoothing out transitions is frequently the goal of lighting designers, to keep the focus on the required affect and hiding from the audience the technical changes that helped to create that affect. Here is Natasha Chivers talking about just that:

> **Natasha Chivers:** I've just done *Boeing Boeing*[4] at the Crucible, Sheffield... [The set] is one open space – and there are a lot of [lighting] cues in it, probably about 35 to 40 in each of the three acts, but most people wouldn't notice. It's funny – I had a friend in during previews, and it was a great success actually, because afterwards she said she'd only seen one [lighting] cue – and the next day I slowed that one down [so the next audience would not see it] (laughter). But it [the lighting design] is moving all the time – just to make sure that they [the performers] always have the right amount of light on their faces – that they're not bleached out or under-lit. I probably could have got away with not doing all those cues, with not doing as much, but I would have been disappointed in myself.

Beyond the look

With increasingly complex lighting rigs, which now sometimes involve video screens and projectors as well as moving lights and other technologies, there are times when 'the look' is no longer an adequate way of thinking about what is going on:

> **Ben Ormerod:** I think, also, there's something else that's happened in lighting design recently, which makes collaboration difficult. As a lighting designer, one has a duty to solve that problem, and that is that lighting has got so complicated. I remember, when I was doing *Zorro*,[5] the director said, 'Could you show me the state when such and such is happening?' and I said, 'No, I can't, because, actually, there are four different cues running at that point, so we'd have to go back to here, run them, and stop.' So that, in the end, was what we did. We put on the rehearsal video for timings and we ran the cues and then we froze the operation and said 'that's the state.' This idea, this convention of lighting states, which comes from manual desks,[6] is becoming less the case.

What Ormerod highlights I think is only the latest version of the potentially difficult choices faced by almost every generation of lighting designers and other theatre makers; simply put, to be led by technology or to lead it.

Technical rehearsals: The process of 'lighting over'

In the early days of theatre lighting design, both in the UK and in North America, it was common to plot each lighting cue state without the performers present on stage. This process often happened overnight. Whilst this practice has not been completely abandoned, advances in technology combined with tighter production schedules and more civilised working practices[7] mean that the lighting for most plays and musicals is largely plotted during the technical rehearsals or tech, with the cast in their performance positions on stage. This is often referred to as 'lighting over'. Scheduled lighting sessions, when they do still happen, are often a time to plot and record potentially useful focus positions for moving lights, and to establish a colour palette for moving lights and LED fixtures.

When lighting design is accomplished with a traditional lighting rig, utilising 45-degree angles from the front of house positions, the effect of a particular combination of lanterns is often quite predictable, particularly for a lighting designer who has lit many productions in that way in that space and spaces very similar. Working in this way, it is not uncommon to plot most of the cue states for a production without the performers on stage. However, as we have already heard, contemporary lighting designers seek to work in a quite different way, frequently placing key fixtures in unique positions, particular to the production and in response to the space created by both the set and the performers (see, for example, Poet in chapter 2). Their light often enters the space from unusual angles, and sometimes crosses the space without leaving a trace – until someone steps into it, and not only in lighting for dance. Trying to create lighting cue states that are specific to the shape and arrangement of bodies in space without those bodies is largely guesswork. Lighting over means that the lighting designer can see the effect that the light has on the performers, and also see how they affect the rest of the light in the space.

For all its advantages – and most of those interviewed much prefer creating lighting cue states with the performers on stage to working on an empty stage – lighting over has the potential to be extremely stressful, especially for the lighting designer and the lighting programmer, both of whom have to be able to work very fast. Here is Bruno Poet on his experience of lighting Cameron Mackintosh's revival of *Miss Saigon*[8] in 2014:

> **Bruno Poet:** For the ten days of actual tech, before the first preview, that's two sessions a day tech-time [the afternoon and evening sessions, with the morning session for technical work on stage] I probably didn't stop talking and Warren Letton [the lighting programmer] probably didn't stop typing for the whole of that time. Probably with only the odd coffee in between. We made one thousand or so cues in that time, and it's exciting. It's being busy.

During the ideal technical rehearsal period, each aspect of the production is rehearsed and then 'run'. For example, if there is a complex scene change involving moving stage trucks and flown elements of the set, for everyone's safety these moves may first be tried out under working light without the performers, then under working light with the performers, then finally under designed light with performers. If there has been no opportunity to at least roughly plot the lighting cue states and changes before the tech, the lighting designer and her team will be under a huge pressure to produce something acceptable very quickly. It is a very stressful business, creating work in public, in front of everyone in the production team:

> **Mark Jonathan:** When I'm teaching directors, I sometimes compare the feeling of how public it is when we bring up our first lights [at the beginning of the technical rehearsal] with standing naked on a table and dancing in front of everybody. So you desperately try and pull and stretch the light, and of course the whole thing is malleable. Maybe you start at the back with the sky and work forward, or you start with your key light – but of course how bright how dark, how warm how cold, how quick how slow – all these things are so variable. And we have no time.[9]

No time, but a lot to be achieved. Ben Ormerod sums up his view of the role of the lighting designer in the tech like this:

> **Ben Ormerod:** In the tech, there's actually not much time to talk about stuff, there's just so much to be done. So, my job, it seems to me, is to give everybody what they want before they ask for it.

A framework of cue points

In the previous chapter most of lighting designers we heard from talked about developing a framework for their design as they read the text and watch rehearsals. A big part of this framework is the technical arrangement of lighting instruments that will be the lighting rig – and the detail of how each instrument will be focused. The need to do this is a given for most if not all lighting designers since, as we have already noted, they require a focused rig to produce designed light on stage. Even when much of the light is coming from moving instruments, there is not generally time to stop and point each one used in every lighting cue state. It is usual to start the tech with a selection of recorded pre-set positions 'in the desk'. Chapter 4 has more on approaches to this.

The other building block most lighting designers want in place at the start of the tech is the framework of cue points. Jon Clark and Bruno Poet both aim to have a set of empty cue states recorded in the lighting console, and their associated cue points noted down in The Book with the DSM or show caller, ready for the start of the technical rehearsal. As Nick Richings mentioned earlier in this chapter, where possible the cue states in the lighting desk will have reference names as well as numbers, and perhaps a short description of the cue point or the planned look.

As we learnt in chapter 2, Jon Clark makes a story-board in the rehearsal room. But it is not a story-board of lighting looks, it is his way of annotating the shapes in the rehearsal room:

> **Moran:** So when you arrive at the tech, by and large you have a cue structure from the story-board you created in rehearsal?
>
> **Jon Clark:** Yes. That story-board always ends up living in the script and I will pretty much notate everything in terms of the action. Then depending on the speed of how it's going I will normally put lighting notes in there, but I will then transfer them to a cue sheet. What I aim to do is have the cue structure by the end of the rehearsal process. [I like] to have the cue structure nailed down. That's not the lighting design, that's just when the production or the play requires something to happen, an event that requires a change, or when I have a feeling something should happen and a rough idea of time and where that cue goes. I will always arrive at that and I will always get all those cues in the book with the DSM at the end of the rehearsal process. That gives me a massive freedom and then I will usually type all of that [cue list into the lighting console, using an] offline [editor] and load that into the lighting desk before we start the tech. I know if I've got that freedom of the structure then the admin involved in putting all that information into the desk during the tech is completely lifted.

To be clear, Clark's cue structure is usually the cue points and some cue times. For some it might include aspects of the particular look they will be aiming for, but only rarely will it include details of the contents of each lighting cue state.

Clark's process aims at being more formally structured by the end of rehearsals than that of some other lighting designers. However, the basic notion of discovering where the cues will go through watching rehearsals is common practice for almost all the lighting designers here. Rick Fisher explains how he develops a cue structure during rehearsals:

> **Rick Fisher:** My process – for want of calling it anything else – is often while I sit in rehearsals, I start to note where I think cue points are going to be. They could be linked to blocking, there could be an emotional

> trigger, or ... I don't always know what that moment is, but I'll just make a list of potential cue points. Around 60% of them end up being cues. There's another 10% of cues I hadn't thought of, or that I get asked for by the director. [Some of those] seem to come out of the blue or left field, or they're a little mental thing – or some staging need that I hadn't ever perceived; I need to hide this or: 'we can't really see somebody behind that door at this moment'. So that list made whilst watching in rehearsal room becomes the skeleton that I start to hang the design from. It's just: 'oh! I feel like there's a change of light here. I feel like there's a build there. I feel like ... something might happen there.' ... I think you just get more sensitive to that with more experience. [You develop an awareness] that something might happen here.

Most of the lighting designers here go into technical rehearsals with clear ideas about where the majority of their cue points will be. One notable exception is Peter Mumford, who we will hear from a little later in this chapter. But even he 'might put little crosses in the script or the music to say: "there's a cue there"'.

How formal or informal the structure of cue points is varies from lighting designer to lighting designer, and also from production to production for each lighting designer. It seems that most lighting designers most of the time go into tech with some idea of where most of the lighting cue points will be.

Making cue states: Creation or re-creation?

When they are lighting over, are lighting designers engaged in a process of spontaneous creation or are they aiming to reproduce on stage something that in some way exists already in their imagination? When I asked this question I had a range of responses. Nick Richings is not alone in firmly declaring that most of the time he is aiming to re-create on stage the looks he has imagined in the rehearsal room:

> **Nick Richings:** [When I'm working on a scripted piece] I normally know when I draw the lamps on the plan what it will look like ... My discovery process is trying to squeeze the lights into the right places during the drawing process. Then in the focus sessions I get very disappointed because it somehow doesn't look like you wanted it to, and then in the plotting session you manage to somehow pull it back. And it's always the same set of highs and lows for me. During the focusing session I do get so miserable. I just think: 'well this isn't going to work' – I get so convinced that it's not going to work. But somehow it does. ...
>
> I always start a plotting session with a spreadsheet – I always know where the cues are going to go.

So Richings arrives in the theatre with a clear idea of where each major cue will go, the cue point, and, based on his extensive experience, a clear idea of what each lantern will do and how they will work together to make each major look. But these are starting points, and there will be a process – working with the director, cast and other creative team members – to discover the eventual finished lighting design. This idea of fully worked out starting points, developed through the rehearsal period, is not at all unique to Richings. Johanna Town understands where Richings is coming from, and sometimes works in that way too, especially when she is constrained by tight production schedules.

> **Johanna Town:** We often don't have enough time [and too often have to] bash it out. ... I think there are some very gifted and crafty people in English lighting design and people don't appreciate what we're turning out so quickly.
>
> That's why you have to have so much of it clearly placed in your head, so I would agree [with Nick Richings] that I do have a plan, sometimes 100% of it, but when I can, [during the tech] my attitude is 'let's go on a journey'. This is my time to play so let's create something.

Even with a more generous production schedule, a lot of the time Town is aiming to re-create pictures she has developed in her head whilst in the rehearsal room:

> **Johanna Town:** I have a picture of what I want it all to look like in my head. I have created these pictures from my script and cue sheet work very early on, I then try and put all these ideas on a plan. There's always more than one thing that is not achievable, that you hope might come out in the wash later. I think I probably achieved 70% of the pictures I want.
>
> That's one of the reasons I don't like fast focus sessions, or focus sessions without all the set, because there are lights that can get switched on and you see something new you weren't expecting – you just never know what little surprise is around the corner that you think, 'That is a brilliant idea. Why didn't I put that on the plan?' and you just keep it there. You tweak it and then it's there and sometimes you may reshape your ideas around that.
>
> I do think that your vision is not always going to be the same as the set designer's and the director's vision. Often you don't realise that until you're halfway through tech session one and you think, 'Okay, I did

explain that to you very clearly.' Then you're trying to work out how to solve this or that problem. It's not that their ideas are wrong but often you have to go away overnight and reform your ideas, you have to be constantly flexible to [everyone's] ideas.

You also see an actor take shape on stage, it is very different than in a rehearsal because you've got the set and everything around them. Not only are they costumed but you've got the scenery in 3D in front of you as opposed to nothing in a rehearsal room. Sometimes in a rehearsal room you go, 'I need to do this, this and this because they're going to be so exposed if I don't.' And then when you get on stage with the set you realise that you don't need to do that at all.

I would say probably only 60–70% of what I want to do gets done, but that doesn't necessarily mean that the other 30% was wrong ... change is a good thing. It's not a bad thing that it's all very flexible.

That's why it's good to have previews – you'll see a show develop throughout [your lighting of] Act 1, then you'll find the key to how to light it in Act 2, and sometimes that leaves you with two different shows in the [lighting] desk, and somehow you've got to get Act 1 back to where you ended up in Act 2; it sounds difficult but it should be a nice journey.

Town's concern highlights a discovery process that many lighting designers talk about. This is Neil Austin explaining the process of adapting to the discoveries made on that 'journey':

Neil Austin: So, the point is you have a certain amount of ideas, a certain amount of weapons in your arsenal that you equip yourself with. 'I know that I'm going to need a little bit of this, and I'd love this, and I'd really like that scene to be this.' But in the moment of tech, the reality of the way the actor's tilting their head and exactly where they are looking and actually what that paint surface on the floor is, and the colour of that costume, and actually what that light really does, compared to what you thought that light might do, that's when you discover and you become very quick at making decisions and changes or completely re-envisioning your idea – on the spur of the moment.

Sometimes you go, 'bugger – that doesn't work!' but you can't just sit there and say to the entire team and all the actors on stage, 'would you mind waiting a moment? I'd just like to go back to my drawing board, and spend another hour having a quick thought about [how to solve this].' You have to say 'well that didn't work! This will – what about that? Hmm – still not doing it. Let's move on for the moment. I know what I need to do. One light here will sort that out [and we can put it in over the next break].'

The whole problem with lighting is that it's indescribable to other people beforehand, and you can only attempt to imagine it for yourself [as the lighting designer]. There are so many variables in the process that even the best of imaginations can't always foresee exactly what it's going to look like until you are in the space. That's why lighting time is so very important.

Austin highlights the pressure on the lighting designer to make a response to things which have not worked as hoped. This can mean that the lighting designer becomes responsible (by choice or through necessity) for coming up with a technical solution. Often, though, especially when working with a larger rig, in a well-equipped theatre, it is more a matter of simply trying something else. Rick Fisher has talked about 'rigging possibilities'. This is how he uses them:

Rick Fisher: I prepare [something like] an old-fashioned artist's palette, with lots of daubs of paint, and they're all ready. And when it comes right down to it – in the heat of [lighting over rehearsals on stage] I throw paint across the stage. And some bits I like, and some bits I don't like – so I take them out. And seeing how it builds – I've got all the angles, colours shapes, flavours, edges – that I might possibly want. Then it's just a question of choosing what sticks.

In the act of 'throwing light at the stage' the lighting designer may discover a better way – a different approach that works better with the evolving dramaturgy of the performance, or as Rick Fisher has called it, the developing language of the piece:

Rick Fisher: Sometimes I get it colossally wrong – then hopefully I get around to remaking that! You know, starting again. Or you start off with a strong idea, but then to make it work you've filled it in so much it no longer is strong ... You think: 'OK – now let's do this all again' and then you go Bang! Bang! Bang! Turn on six lights, turn off three of them, and then you've got a great looking state.

Moran: So is this the sort of thing you mean when you talk about establishing a language for the piece?

Rick Fisher: Well I think it's about finding out which lights and colours, which types of light work. And it's surrounding people with the right quality of light and the right quality of darkness – what we are not lighting. Let's not forget that the darkness should be just as designed as the light. How that helps us connect with the people by lighting them and not what's around them. Removing distraction – that helps to form the right light. I can't always describe it but I know I know it when I see it. I think that as you get more experienced, that's what you develop a taste for. [It's about] what feels right.

Moran: Could it be as mundane as saying: 'this quality is day time, this quality is night time?'

Rick Fisher: It can be. What fascinates me is how we can believe it's night time in so many different ways. Or we can believe that someone is outside in so many different ways. It can be a cliché as, we are going outside so the floor gets all sort of dappled, or the air gets sort of dappled. Or it could be that somehow the colour just changes in some way or the volume of light there is saying that we're outside. But I don't think about it, I just do it. I can talk about it, but it's trusting your instincts and developing your instincts.

Group improvisation in the tech

It is important to reiterate that the creative work of the lighting designer in the tech does not happen in isolation. Different people used different analogies to describe what's going on when the tech is going well. Rick Fisher throwing "paint" at the stage, for example.

Natasha Chivers used the analogy of making music together, defining herself more often than not as a rhythm player supporting lead melody lines. It's an analogy that appeals to several other lighting designers, as we will see later in the book:

Natasha Chivers: [Early in my career, I was working closely with Rufus Norris and Katrina Lindsay as co-artistic directors] and it taught me that I'm not a lead singer – I'm a drummer, or a bass guitarist – and I'm happy with that. I liked being in the band and I knew that I wanted a part of me – and a significant part of me – to appear on that stage. And I had an instinct...

I want to make theatre and I had a really strong sense that I wanted what I do to be really central to that production.

Moran: It's interesting that the instruments you choose in your analogy are drums and bass. So you're talking about rhythm and pace ... that sense of having a responsibility for how the audience get the speed of what is going on.

Natasha Chivers: Yes – and I think that's often forgotten about. I mean there's the colour – but you are the editor of it as well – and I think theatre lighting design is extraordinary, because it's not just about 100+ images that operate separately, but they run into each other in a way that's appealing – then you're setting the entire pace, and you're pulling focus so you're telling the audience which way to look. There are so many complex things going on that even most theatre people just don't get. And in a way we are our own worst enemies because through our work we do explain so much of what is going on to the audience – but most people don't get that.

Perhaps in the same way that most people find it easier to recognise the contribution of a lead vocalist or instrumentalist than the importance to the overall sound of a great drummer or bass player just doing their job?

Moran: So if we return to your drummer / bass player thing ... is this jazz improvisation? You know the basic structure, say the chords and the rhythm patterns, and you have to try and work out...

Natasha Chivers: Yes – and when you need to be more prominent and less prominent and when there's a pause, and when it takes off – yes, it is like that, it is coming from somewhere that isn't totally conscious.

Chivers is talking about making responses in the moment, informed by an understanding of the evolving structure of her lighting design, and its part in the broader structures of the whole piece. That is part of something she calls instinct. Some of what Fisher is talking about could be described as a highly developed way of looking at the stage and the construction of the stage picture – what the great American lighting designer and educator Jennifer Tipton refers to as the need for the lighting designer to be a 'super-audience'.[10] Some of this instinct could be called experience – knowing what has worked in a range of previously encountered situations. But, as I think Chivers' analogy makes clearer, some of the instinct relies on being able to read and understand the structures of this particular piece – both the inherent ones, found in the text or revealed in the rehearsal room, and those that are developed by the team (including the performers) in the tech.

The lighting designer's instinct

Another part of the instinct manifests itself when the right bit of kit is in the right place to make the right response in that moment. This skill is informed by an ability to visualise the lit space long before the tech – in the neutrally lit and blandly furnished rehearsal room, for example. Some LDs develop this skill through years of practice. Ben Ormerod is one who had it right from the beginning, and he talks here about how it works for him:

> **Ben Ormerod:** [Y]ou have to play it through in your head, that's the thing you see. You watch a run-through with all these elements and you go, 'Does it all work?' Sometimes it doesn't and you have to think again. This is another reason for being in rehearsals a lot because by the time you get to the run-through you want to know what it is you're going to do, so you can try it out in the run-through [in your head].
>
> You try it out with your third eye, your inner eye. You see really what happens with me and you ask about a process; this is the process bit, is that whilst I'm watching rehearsals I'm seeing the show lit. I mean I'm not thinking, I'm not making that happen it's just that's what happens. I'll sit in rehearsals, they start acting and it starts being lit. So then my job as a lighting designer is to work out how to do what I've just seen ... that was something I could do more or less from the beginning ... What I brought to the table from the start was the ability to just sit and watch actors and for the light to appear on them like that.

What Ormerod has been able to do from the beginning, others have found ways to achieve too. How else could Richings, Town, Austin or many others have such a clear idea of their starting points before a light has even been focused, let alone become part of a lighting cue state? For anyone who does not begin with Ormerod's useful talent, developing that 'third ... inner eye' is surely part of developing the lighting designer's instinct. Just like almost all creative practitioners, the lighting designers here have a history of practice, mostly self-taught or learning by working with others, and developing their instinct. Bruno Poet spoke about the 60-odd shows he lit whilst studying for his geography degree at Oxford being a great way to experiment and develop an instinct. Lucy Carter spoke about studying Peter Mumford and writing a dissertation on his work, and of working as a technician at The Place in London, lighting for lots of choreographers in training. Jon Clark spoke about how he consciously developed some of the qualities that I'm linking to instinct in his early career working as a re-lighter for more established lighting designers:

> **Jon Clark:** [I]n my twenties [I was] more or less lighting what I could when I could but also taking quite a broad range of technical re-light[11] work as well. Your opinions start to form more clearly, especially about the right light I suppose, and the process. How you start making that right light and the choices that you make from re-creating other people's work on your own, but also creating their work with them for the first time – being behind a light and focusing it. The mechanics of how you achieve things quickly and efficiently by copying and observing other people, but also the choices that are made in terms of light and where it comes from, how you build a picture, how you build a lighting design for a piece.

Now very much an established lighting designer in his own right, Clark's take on developing a design for a production is in some ways an extension of Fisher's notion of a language for the production. Where Fisher says 'I don't really talk about it,' Clark consciously sets rules for himself, constraining what he allows himself to do with his rig at any particular moment:

> **Jon Clark:** I try to be very clear that it's a design for that production and therefore I try to live within the rules that I then impose on it, but it has to come from the centre of where the production is coming from. From the play and work from that point out, so there can be some things that I won't allow myself to do within that frame – it should be one design for the whole piece. Everything should be coherent within that world. It's a world and a choice and a decision you're making about how you want to explore and interpret that production as opposed to serving and breaking down scene-by-scene requirements for that production. Finding a language for the whole piece rather than saying 'Scene 1 is daylight so I'll stick ten lights and light this as daylight' and I'll start again for scene 2. [I don't want to end up with] ten or eleven individual lighting designs, which may all be aesthetically appropriate for each scene. I'd rather take an opinion on an entire production,and design a *world* that explores that and deals with that as a piece of design and a quality and direction of light within that. ...
>
> It's hard to generalise! I like, if I can, for the structure of the rig to inform the space as well, for that to be present, so that your interpretation of the piece can work within a very defined *world*.

James Farncombe is of the same generation as Jon Clark. He too has spent years building up his lighting and general theatre instincts and spends time in rehearsal building an instinct for each particular production. Unlike Richings and Town, however, he does not like to go into the technical rehearsals with too clear an idea of what each state will look like. This extract starts by framing the question in terms of fine art practice:

> **Moran:** There's a thing about producing fine art where a discussion has been how much of what the artist is producing is an image that pre-exists in their own head and how much is it an experiment where you discover things along the way.
>
> **James Farncombe:** That's interesting. I think I actively try to avoid getting too married to ideas. I consider the drawing of the [lighting rig] to be the first of several steps. There's a huge amount of preparation you can do. I've stopped putting too much store by that paper rig design. I think a lot of the technical information for speedy fit-up has to be there, but thereafter, everything is up for grabs.

Farncombe, like Johanna Town earlier, goes on to be enthusiastic about making use of the happy accident – the light from a fixture that skims interestingly across a surface on its way to being pointed to do a different job:

> **James Farncombe:** I know that there's only so much you can anticipate, both negative and positive. Say someone's climbed a ladder at my behest to focus a lamp and as they get it into position it's skimmed across a wall, you go, 'that was brilliant. Put that back there.'
>
> OK, now we've got to find something to do the original job but there's a bunch of stuff I didn't know was going to happen and so you're actively thinking on your feet and responding to those things and incorporating them if you think they're appropriate.
>
> I suppose that extends into the relationship between director and designer as well because they should be feeling that although they've got an idea as to what it's going to be, when we get into the space there's going to be other things that we discover. You enter into it, the process that is, with the spirit of adventure. You set out to discover stuff.
>
> If you're chasing that image in your mind's eye, for a start it might differ from the [image in the] mind's eye of the designer, because it's so hard to communicate that information isn't it? Especially when we're talking about something as ephemeral as lighting. So it's an experiment in that I know there are basic things that will work, but there's also a higher echelon of stuff that I'm looking forward to discovering whenever I start a show.
>
> **Moran:** To what extent have you already decided where the cues go?

James Farncombe: Again I think there's probably a skeleton structure. I will sit in the traditional fashion in the rehearsal room and mark where I think cues go. Invariably they move. There are certain cues that you know are always going to be there and depending on the structure of the show I suppose.

There are certain cues that you know are going to be there and others that are a little bit more floating around and settle eventually as time goes by; I have a framework that I work to – for sure – but that moves/changes.

Peter Mumford: starting with nothing?

Unlike the lighting designers we have already heard from in this chapter, Peter Mumford does not begin with a structured cue list, or (it appears) a clear notion of the looks, or the contents of the key lighting cue states. Perhaps this practice stems from Mumford's Art School background – uniquely among the lighting designers here, he had a formal training as a stage designer at the Central School of Art and Design. Mumford also works almost exclusively at a level where he can expect or demand the best lighting programmers to be working alongside him. It is perhaps because of this that he feels able to respond so freely in the tech, unencumbered by a predetermined structure of cue-points. Perhaps many of those whose process includes starting the tech with at least a skeleton structure do so in part to give themselves the head start that Mumford gains from collaborating with top programmers.

It is clear from this extract, though, that even Mumford begins the tech with some structured ideas about both where the cues will go and what the key looks will be:

Peter Mumford: The way I approach a lighting design, the technical bit, the drawing the plan, the looking at the model box and the costume designs, going into rehearsals and watching rehearsals, I regard as a palette to work with. So although I take notes, and I might put a little cross in the script or in the music to say there's a cue there, I'll notate a piece, going cue point cue point cue point, but not very scientifically at all – just where I know, or I think, some kind of change needs to occur, which might be a move or a change in the music or whatever. But the rig for me – the rig plan that I've drawn up and asked for – is a palette to work with. So I don't do any kind of numbers [prior to the technical rehearsals]. If the DSM comes up to say: 'how many cues in the show?' I say 'I don't know'. And I don't – but I have a rough idea, I have a sense of the structure, I have a sense of the kind of look, but not usually in a totally precise or scientific way. Because once you get in there and start working, you know immediately if your palette is going to work.

Moran: Is that at the focus or...

Peter Mumford: Well it's after the focus – you focus and then you find out if the colour and the focus you've done is right. I mean obviously you've thought about it. If you get it wrong you're f****d basically. But once that's there, then I do allow quite a lot of flexibility. I wouldn't say I was making it up as I went along, I've got certain specifics. But I will – if I've got a number of moving lights in the rig, and let's say it's a kind of standard set ... I will get my programmer to get every mover to be able to go to every piece of furniture, and set up those groups to start with so that we can do sofa, chair, window, backlight, front light and so on, so that I can work quite fluidly in terms of re-arranging and pulling focus – and some of that you can't see until you're working with the actors [during the tech].

Moran: There is a great book by Gilles Deleuze, a French philosopher, based on interviews with [the artist] Francis Bacon. At one point they are talking about the thought process behind a brush stroke, knowing where it's going to start but not necessarily where it is going to finish (Deleuze, 1981/2003).

Peter Mumford: I think that's a very good description.

Natasha Chivers: making it up as you go along

Chivers talks here about a production process that finished up with thoroughly composed images. Despite the defined starting point, the text of Shakespeare's *Macbeth*, the piece did not follow the formal structure of the play-text, and was effectively a one-man show. She talks here about the joy of making work that depended on high-level instinctive response and tapping into the creative empathy present in the room:

Natasha Chivers: I loved my *Macbeth*.[12] I can honestly say that almost every time the lights changed I loved what I saw on stage. Every single image was composed thoroughly. I thought the set was amazing so that really helped me being able to do detail, and light with precision.

Moran: And was that story-boarded in your head before you started lighting it?

Natasha Chivers: So one of the reasons that I'm most proud of it is that when I went up to Glasgow to see it for the first time, so I'd just come

> back from China (from a research trip...) – and I'd also just split up with someone – so I emotionally I was a bit on the edge, so I'm particularly proud of having sat in that chair [at the production desk] for a whole week, when for many reasons I'd really rather not have been there, and achieved under those circumstances what I achieved. And I think it is a testament to what you have up your sleeve, because no – I didn't actually.
>
> The space was huge – the Tramway (in Glasgow) and I think the set was something like 18 metres wide. And there was only one man – Alan Cumming. So there is this huge green mental ward-like cell and a little observation room at the back. Massive, and quite bleak as well. One man on there and ... the lighting was like a follow-spot – the whole place was lit but the focus was always on Alan, millions of cues, and you got to know the blocking really well.
>
> It was about a man with severe mental illness. He arrives – the character – and he's done something, and we don't know what, but this person plays out Macbeth as part of his neurosis. ...
>
> You literally make it up as you're going along, and that's what's so good about it – about this job. Because the thing was that although it was based on Macbeth, it was all hacked around, and we didn't really know the order, and because it was about madness within an institution, for instance they didn't know what the Heath was going to be – and Alan came up with doing the Lady Macbeth speech in a bath. So none of us knew really how it was going to work, but you know [as a lighting designer] you see several run-throughs before you've got to do it.
>
> I think videoing rehearsals is a life saver these days. Just so you know where you're going.
>
> But sometimes the instinct is so strong ... I remember with the *Macbeth*, you might not even be entirely sure what the plot is at that point, but you're still able to do it. It's weird. So a scene might have been cut up and you're not quite sure what's going on, but you get into a kind of zone there, you pick up on what's going on in the room and you just follow it.

This is a good example of a lighting designer tapping into an empathy within the room where exciting work is being created: Chivers' response coming from the same creative well as all the other theatre makers in the room (including the key solo performer), together making something greater than the sum of its parts. And thinking back to reason two in the first chapter, where the seams between its parts can't easily be seen.

By the seat of their pants!

Although there can often be a creative 'journey of discovery' during the technical rehearsals, more often than not at this time many of the people involved are primarily concerned with getting the show safely into a repeatable state. Many things are happening at once in a tech and it is very rarely time for in-depth conversation and quiet contemplation. It seems that for most of those interviewed, for the work they judge their best, something has developed over the preceding weeks of rehearsals – or years of working together – that allows a high level of empathy to exist in the creative crucible of the tech.

We heard in chapter 2 how Paule Constable wants to be sufficiently inside the piece and the director's vision of it to be able to 'rough something in' while the director keeps working on other things. Here she gets passionate about that feeling of creating in the moment; something that happens a lot when a lighting designer is lighting over a rehearsal on stage:

> **Paule Constable:** I'm constantly flying by the seat of my pants!
>
> **Moran:** And is that part of why you do it? [the job of lighting designer]
>
> **Paule Constable:** Yes of course it is. We [theatre lighting designers] are all addicted to that slightly aren't we? And when that's going well it's amazing.

This is not in any way to advocate a totally unstructured approach. For Constable and many others, every response has to be made with a deep knowledge and understanding of the piece:

> **Paule Constable:** I think having an intellectual approach ... a practical, critical rigour underneath my thinking has been really useful in that way. My obsession in my work is not doing arbitrary – do anything as long as you've got a reason. As long as you've got something to hang it on – never just because 'I think it looks great'. Because that's not good enough.
>
> And you can see kids doing it. The first time they do a cue and it's orange and blue, and they go: 'mmmm! That's great!' and you go: 'well yes – yes you're right, but there's more to it than that. Cool and warm? Ask yourself why it works? Look at other places where you see that resonance. Look at Japanese art.'

This last statement perhaps speaks to something mutually understood but rarely mentioned – why does it so often take so long to become an accomplished theatre lighting designer? Even with formal or informal training? Few of the people here made any kind of name for themselves in the first ten or so years of practising. I have already mentioned different ways in which the lighting designers here took time to develop what they have called their instincts. It appears that this is not something that can be done quickly. Another part of the answer may well be the sheer range of experiences and influences that need to be absorbed and internalised ready to be put to use at a moment's notice, and the confidence to make choices under pressure during a tech.

Later Constable talked in detail about her own process of discovery, through rehearsals in the rehearsal room and then on stage during technical rehearsals and on into the preview period:

> **Moran:** So you mentioned story-board there; do you always story-board?
>
> **Paule Constable:** I tend to write the story of the show. I don't sketch it because I think that's too difficult. But I write stories for myself – this is this and that is that [for my lighting] – and the muscle of a scene. And those stories start right back from the beginning of the process. So in a way ... when we do a run I take that story and by then I'll be able to break it down into cues, and what's going to happen in each cue. And again, that sounds sterile, but it can all change – it just gives you building blocks.
>
> **Moran:** So to what extent are you aiming to build on stage something that pre-exists in your mind already?
>
> **Paule Constable:** What I quite often do, partly again because of the opera thing, is something like: Scene 1, OK, so I know that the downstage right bedside light is on. There's a bit of light behind that door, and there's a flat greyness in the room. So I'll say [to the programmer]: 'OK, put that practical at 60%, put the light behind the door at 80% and put the top light grey – 728 [LEE Filters Steel Green] cover – at 30%,' and then I'll step away. Put in the next 20 cues as that, and then put a block in.[13] So then I can pick or choose or whatever. So that's kind of in my head, and quite often that's wrong. But even by turning it on and looking at it and just going: 'oh – shit! OK' [you get closer to what will work] (laugh).
>
> [Away from the stage] you do do mental experiments with that kind of feel thing as well don't you? As you're sitting watching a rehearsal you're thinking: well it is all going to be keyed from that side (say because that's where the window is) but actually they are all playing it to over there, so I need to switch that on. Or I need to say to the director, 'actually ... if they were in a room with the window there, they would play it over there. Would that work or do we need to look at a different approach?' ...

Moran: How long do you think it took you to be able to see in your head in that kind of detail?

Paule Constable: Yes, a long time. But I think it's amazing, if you spend time in rehearsals, your intuitive response to something gets formed and articulated by an actual response to it, so you turn it round and you have a sense of it and you think it's going to be this or that and as you feed rehearsals into that sense of something, the dynamic of it is actually informed by what you are seeing unfold in front of you. Be it devised or whatever it is. ... It's a really dangerous process, but it's an informed one.

Moran: And it's exciting too – discovering the space that works for each moment of the piece...

Paule Constable: Yes, yes. And the rhythm of things – you know something is doing this and then suddenly – it takes shape, and you think wow! Suddenly there's a sense of something shifting and landing, that it would be gorgeous to link into.

Moran: And does that start to inform your cue structure as well?

Paule Constable: Yes, absolutely.

Moran: So when you put in those next 20 cues you were talking about, you have an idea of cue points? And cue times too?

Paule Constable: Cue points, absolutely. I always give them to the DSM before we hit the theatre. I'm a bit rubbish at cue times, in that my first attempt - unless something is really obvious - I'll just put a default time of 9 or 7 or whatever it is, so it just happens, so I can go: 'OK'. But I know when something is quick and when it's slow, and when it needs to land.

Sometimes that thing of how quickly or slowly a change can be sustained by a space is quite difficult to tell until you see how the textures and surfaces are responding to light. You know how when something is very active, and you shift [the intensity of the light hitting] it by 10% and you go: 'woo!' and you realise that you need to sneak it in a bit more slowly. And sometimes you've got 50 chorus on stage and you can splatter light all over the place and no one even notices! (laugh)

As intellectual as Constable's approach is, what is most important is what it looks like on the stage with the performers. She went on to talk about some moments when that look has been inspired or perhaps driven by more conceptual concerns:

> **Moran:** The business of 'light motif' is quite tricky – but light as metaphor, you have used really powerfully at times – just wondering about how you might go about selling that idea to the rest of the creative team? ... or do you just present it as a thing on stage?
>
> **Paule Constable:** I think sometimes ... it's a very tricky question isn't it because in a way ... I'm trying to think of an example that I would hang that on. Maybe the Schiller plays [directed by Michael Grandage] – *Don Carlos*[14] Ah! That's interesting ... let's take another Schiller – *Luise Miller*.[15] If I'm talking about it in really, really mundane terms, there are three sorts of scene in *Luise Miller* – there's her home life, there's the court life – the oppressing political court life, and there's the sexual court life. And you think: 'well what are the signposts the audience need to sustain that?' So you kind of go – well the landscape of her home life has to be simpler than the court life. And the court life has to be oppressive, and dominant. And then there's the sexual life, with the Alex Kingston character [Lady Milford]. So the metaphor actually starts within my way of thinking about those [three sorts of scene].
>
> Schiller is brilliant – like Shakespeare, like Mozart, like Chekhov – he is using the light as a metaphor in his writing. It's very literally day or night.
>
> It's often the same in Chekhov's plays. *Uncle Vanya* is a descent into darkness. And that's a metaphor, but he actually does it in the writing. Each scene is darker and darker and darker, to the point where – at the end Sonya is talking about having to light candles, to hold the darkness at bay as Vanya is tipping into despair about death.
>
> And it comes back to having been an English [literature] student – the clue is to trust the writing. Now not all playwrights are as great as Chekhov, or as aware, but you can impose that sort of structure, if you have discovered it. You can create that – but so often it's in there already. It's slightly with an external hand that you go: 'I'm going to heighten this, am going to do that,' but the rules are that it's about the story, and a response to the story.
>
> **Moran:** When you are talking about this to directors, certainly a lot of UK directors are – if not English graduates then very well versed in analysing text...
>
> **Paule Constable:** That's true, but it's very rare that I would have that kind of conversation with a director. With someone like Michael Grandage

> I'd make something, and I'd go: 'I've decided that these are the different spaces'. And then it is like you turn up the gas don't you – you kind of separate those things out more. And that in a way is what I'm quietly doing in the background.
>
> If someone comes up to me and says: 'I want this moment to be a coup de théâtre', I kind of go: 'well I want us to have got there because the writing has got us there already, rather than me imposing it'. I think my work is quite self-conscious, but I think for light to have too much of an external hand is really dangerous. It has to be informed. Again it goes back to informed choices, doesn't it? So I'm slightly scared of the metaphor, though I love it when you can use all those things and land a metaphor using light in a way that feels right. It's when it feels like you're forcing an audience that it gets too much – you know: 'I want you to feel this – now!' (laughs)

Earlier in the interview Constable acknowledged her debt to Ormerod, and certainly here there are some similarities to his approach. Perhaps the final paragraph above illustrates as well as any words can the balancing act that Constable – like many others – is attempting in making her work. For her, a key goal is to create an intellectually informed aesthetic, that comes from the text and helps to tell the story, but that still allows enough space for the audience to create its own response rather than being told what to feel.

Many of the lighting designers here would be uncomfortable with the word 'intellectual'. I suggest, however, that the understanding of the piece that Constable calls 'intellectual' is not far away from the 'instinctive' response to text of other lighting designers – and instinct that comes from years of watching and making text-based drama, opera and dance. I have heard several lighting designers, who would shy away from being called intellectual, making deep and pertinent points relating to the text, inside and outside of a rehearsal room or tech. It seems to me that lighting designers who successfully do text-based work acquire an understanding of how a text works that would exceed many Masters-level students of dramatic literature, but that few of them are comfortable putting that into words. Despite this reticence among lighting designers, I will name part of what they call their 'instinct' as an intellectual understanding of the piece.

Structure taken to extremes, for a particular purpose

To conclude this chapter I'm turning again to Ben Ormerod. Here he describes an extreme case of a structured approach to lighting design. It was made in response to the tightest of production constraints (in terms of no time and very little budget) and high aesthetic expectations. This came out of a conversation about Wagner's use of the leitmotif, particularly in his Ring Cycle, and how that informs Ormerod's approach to structuring his lighting design response to music. The staged concert of Harrison Birtwistle's opera *The Mask of Orpheus* referred to here, took place at the Royal Festival Hall in 1996:

> **Ben Ormerod:** I think the thing that really made the huge difference to the way I work was lighting *The Mask of Orpheus*. Harry was doing one of those weeks at the Southbank that they do every year, where an artist curates, and one of the things that he wanted to do was a performance of *The Mask of Orpheus*. Originally, it was just going to be a concert performance but he wanted it to be more than that.
>
> *The Mask of Orpheus* is a huge learn, so you couldn't ask the singers not to sing from scores without paying them for months of preparation. The director, Stephen Langridge, had the idea of employing actors to play the roles, putting the singers in the orchestra and then Alison Chitty designed two stages, one in front of the orchestra for the actors and one behind for the dancers.
>
> So, the Cholmondeleys and the Featherstonehaughs were dancing on this platform behind the stage and then actors were playing the roles on the front stage, and the singers were in the orchestra. It was a completely brilliant idea. Then we got money from the National Studio to rehearse the actors for four weeks, to tapes of the original ENO production[16] and then we took it all into the Festival Hall.
>
> Now, this was the schedule. Our show was going to be on the Sunday night and there was a concert on the Saturday night. When the concert finished on Saturday night, we put in the truss; we rigged the lights – no moving lights and no scrollers in those days. I came in at four in the morning to focus, at ten in the morning we had the dress rehearsal, then everybody had the afternoon off, except us. I had to do my notes and then in the evening we had the show. No plotting time, no tech time. Rig, focus, dress. So I mapped out every single cue with timings a week before we opened and my young associate/production electrician, Neil Austin, and I went in and we sat at the Galaxy [lighting console] for two hours and just typed the whole thing in. Then we rigged it and we focused it and we watched it.
>
> I don't know whether Harry [Birtwistle] would approve of me calling his work 'leitmotific', but, there's no doubt that there is a great deal of repeated

material in *The Mask of Orpheus*. Birtwistle has somewhere described this as 'as if you're walking around a town that you don't know and you keep coming to this market square, but every time you hit it, you come down a different street, so you see it from a different angle.' In order to be able to programme the show without looking at it I created a series of images that were associated with musical material. There's an enormous amount of material in *The Mask of Orpheus* and you have to pick and choose what it is that you're going to light. I think I found something like 15 images, musical images, that I felt would be key to the lighting.

I don't know how you choose those images, but it seemed obvious to me which they were. I had something called 'Ceremony,' I had something called 'Fire,' I had something called 'Water,' I had something called 'Corridors,' and so on and so forth. They were all associated with things that I had heard in the score, and that I recognised on the page were coming back again and again and again. Once I'd created those 15 images, I would play them against each other in the same way that Harry did in the score; sometimes several images might be on stage at the same time. These were the design; they weren't in addition to some kind of 'cover,' they were all there was.

So we ran it and it was fine. I think everybody was very pleased that I'd managed to get so much done in the amount of time that I'd had. But, it wasn't good enough, obviously, I mean, we'd lit the whole thing blind, how could it be?

I then had notes on the elements, so I would say, 'Right, okay, so fire, I achieved it in the wrong way, it's too muddy, I know exactly how to achieve it now, I need to achieve it this way. That's a refocus and a little bit of a re-rig and that's that. Ceremony works. Corridors, they don't work, I made a mistake with that, I'll take the gobo out and I'll do it with shutters instead,' and so on and so forth. I only had ten notes, you see, because they were about the elements rather than specific cues, but they ran through the entire piece. Then, in the afternoon, we did those ten notes and then I had to re-plot sections, because I'd changed some circuits around, but we did a little bit of patching to get things working.

That evening, we had a completely different looking show; it was transformed. But although it looked completely different, none of the cue positions had changed, none of the timings had changed. Those were all right. Structurally, nothing had changed, but the show looked completely different.

I'd already used these techniques, working on lighting another piece by Harry, *Punch and Judy*, but I really went wholesale into them this time, and I realised that this was the way for me to light. It comes from the Wagner, but also from studying Brahms and seeing how he will begin the first symphony with tiny amounts of material, two or three-note cells, and elaborate them into all the material for the first movement; or Haydn, in his string quartets, where he'll take one idea and create an entire string quartet out of it.

Moran: Maybe stretching, or going in the wrong direction with this musical analogy, but with that Harry piece, is [what you did between the rehearsal and the performance] the analogue of re-orchestrating something?

Ben Ormerod: Yes, you mean in terms of the notes session?

Moran: Yes.

Ben Ormerod: Yes, it is, it's absolutely that, I suppose it would be like having composed it and written it and then you hear it and you realise that actually, the way it's voiced is wrong. It's like re-voicing. ... [But] I think the inner ear of the composer is more reliable than the inner eye of the lighting designer.

4 Collaborations

Theatre lighting designers working in the UK today are very aware that they cannot make work on their own. Everyone interviewed here emphasised the importance they place on collaboration in the creating of a lighting design.

The people lighting designers collaborate with fall into two camps – fellow creative team members (including in many instances performers) and those who provide technical and other support, directly and indirectly, for the realisation of lighting designers' work. One of the skills required of theatre lighting designers today is to be able to talk to both these groups of collaborators in a language each understands and respects. While most directors and set designers are not interested in talking about the technical detail of how a particular volume of light has been created, most technicians expect precise specifications for the equipment involved, and the lighting designer must master communication with both groups of collaborators.

This dependence on collaboration – both in coming up with creative ideas and in their technical realisation – makes some lighting designers nervous about calling what they do *art*, and I'll return to that matter later. In this chapter, however, the focus is on creative collaboration. In any collaboration, arts journalists and others are prone to ask the question, 'whose great idea was that? Who is the *creator* of this work?' Setting aside for the moment the possibility that this question implies an, at best, unhelpful way of looking at collaborative work, how do lighting designers respond to this?

> **Moran:** Do you feel disappointed if the end result turns out to have more of the director's ideas than your ideas?
>
> **Nick Richings:** I don't really. ... Most of the time, if you're lucky, the visions run in parallel, so it's really not a problem. But if you're involved in work where you're not happy where it's going, you have to pick out two or three things and stick out for those, and fight your corner for those and give in on the things that don't really matter, or matter less. I think that's true in negotiation for anything.
>
> **Moran:** What are likely to be the things that really matter for you?
>
> **Nick Richings:** Anything that looks like bad lighting. ...

Richings here is concerned with protecting his professional reputation, and the possibility of future work and future income. It is good to be reminded from time to time that for all the lighting designers here their primary or sole income is the fees (and occasional royalties) gained from lighting design. Having said that, it is clear that these lighting designers work hard to ensure that Richings' expectation – 'most of the time ... the visions run in parallel' is what happens. Jon Clark goes into some detail about how he goes about that:

> **Jon Clark:** Some directors want to sit down and talk a lot which is great and there are others who can't or don't want to invest the time in doing that. [I've worked with] some really big directors and it has been quite hard to have a detailed conversation about anything at all. ... Clearly it's quite hard to have a conversation if you don't really have a sense of it [the piece]. If you put yourself in the [rehearsal] room and you're experiencing [the piece] then you start to have responses to what you're seeing. Then you can ask more pertinent questions I suppose.
>
> I guess because you're there you're kind of investing in it and that often rubs off as well. So then you can find yourself in a positon to kind of crack open the production and the play and your process and also you can inform the director's mind from a lighting perspective. Just start that dialogue, and so when you end up in the technical rehearsal, at least you've got key points and ways to get into the production and the play. It might be about how a scene is informed [by light] or how someone is travelling.
>
> The movement of light [in time, through the piece] I think is also crucial. I want [the lighting to respond to] the whole arc of the piece. Aesthetically I want it to look like the lighting belongs to the production, that it's a statement and a choice that has been made for that production.

Johanna Town talks here about how she works to keep 'the visions running in parallel' during the tech:

> **Johanna Town:** You have to be engaged, I think you have to be able to have a conversation. You have to be able to spin on a sixpence in a tech and go, 'Okay, that is not what I thought you meant, I'm just going to act on that, I have to act on it, I have to keep going.'
>
> You're constantly hearing stuff like, 'Wouldn't this be an amazing idea to do this whole scene change this way?' and you have to go, 'Yes, you're right. I know we didn't talk about it but yes, you're right, and let's just action it. Let's be creative.'

> So I do think we have to be incredibly on the ball and not stuck in any rut. You have to have absolutely open thoughts the whole time, which is why I was saying you only ever do 70% of what was in your head because being in an auditorium, you are creating something, and you have to be creative at that point, not back in your office.
>
> The whole talking to people, you have to engage with the person you're working with, and I think it varies from director to director what you talk about. When I wasn't confident about talking about my work, I'd take an awful lot of picture books with me and show art and artists and photographs and sometimes architecture, sometimes landscape ... just to say this is the mood that I'm trying to get across here. I'll talk about films and music, all of those sort of things, they're usually very visual things rather than intellectually book-y things.

For some of the well-established lighting designers interviewed here, the relationship they have with certain directors goes back over many productions. Especially when working with the older generation of theatre directors, some collaborations appear to happen almost without any words. Paul Pyant is typically self-effacing in the way he describes working with two knights of British theatre:

> **Paul Pyant:** I have worked with particular directors because, well I tick a box. For Trevor Nunn or for Peter Hall or anyone like that. With me on board, as far as they are concerned, that ticks a box in their mind and it'll happen. They know they don't have to have any conversation with me or whatever 'cos you'll just do what's right...' which is a very trusting relationship I suppose. They very rarely come in with any thought. They might say, 'Oh it's a bit dark over in that corner.' But that's an old-fashioned way [of working].

There are times when the director or choreographer can over-estimate this however, perhaps forgetting that in a successful collaboration there needs to be some communication:

> **Mark Jonathan:** I think it was *Protecting Veil*, maybe it was something else, but it wasn't something that's completely narrative, the choreographer [David Bintley] said, 'You always know what to do.' I said, 'I only know what to do because you talk to me. I need to know what's in your head and what's going on.' (Laughter)
>
> [From his point of view] the relationship was so comfortable, it was like, 'I don't even need to talk to you.'

Mark Jonathan went on to talk more broadly about the lighting designer's need to be alert to different ways of working with every new creative team:

> **Mark Jonathan:** We have to be chameleons as well in how we relate to our other creative collaborators. I've said it already; how dominant, positive or tacit we are, that varies with every creative relationship.

Earlier we heard from Paule Constable about how important it is for her to get involved early in the process. Here she expands on how she goes about informing those early decisions, beginning a collaboration in which she can do good work:

> **Moran:** You have also said – and lots of others have said this too – that it is difficult to talk to anyone who is not a lighting designer about light. Particularly directors. Given that, what are your ways to do that in those early conversations?
>
> **Paule Constable:** I think a major part of our job is alerting them to the repercussions of these early design decisions. I think there's a lot about trying to imagine the repercussions for them! ... There are now directors that have worked with me a lot, so we've got a language here that we can relate to things that we know about, things that we've experienced together in the past. If it is a director I haven't worked with before it's different – and I've just had this in a production of *Marriage of Figaro* where I had to say to this director: 'if we continue in this direction, what it means is that the piece becomes very installational. It's quite cold – I can give you one thing for each scene, and that's what it will be. It won't move.' Well I did play devil's advocate, which I can because I know the piece very well – 'it won't be about light and air. It will be about being in a space, being stagnant and still. And if that's OK, then let's keep going in this direction. If it's not then we need to have this conversation between us [the whole creative team] sooner rather than later.' So you couch it in terms that mean something to how they are going to direct the show.
>
> Which is fine if you are working with a director that can anticipate how they are going to direct the show, but if you're not, it can become (laughs) tricky!

It may surprise some to hear that there are directors who begin rehearsals not really knowing how they are going to direct a show, and on a larger scale it is not that common. However, there is a whole tradition in the UK of making work in a collaborative and devised way, where everyone agrees to go on a journey to discover how the text works in this space, at this time with these talents in the room. Constable – who has worked extensively in this style, with Complicite and others, is not arguing against this way of working, but rather finding ways to point out how early decisions can put huge constraints on the production – constraints that, as we have said earlier, it takes a particular type of technical and design experience to anticipate.

The notion of the need to establish trust is common, and often comes, as Constable and others have said, from having done a lot of work together. But there are other ways of establishing trust. Johanna Town asks her directors questions that some might find odd for a lighting designer to ask since they have nothing to do with light:

> **Johanna Town:** I've just worked with Blanche [McIntyre, director of *Tonight at 8:30*]. [At the start of the process] we'd had a chat with the set designer, we'd talked about reference material and things to read and all the usual stuff, and I'd gone away and started my research. Then just before I handed the [lighting] plan in [to the technicians responsible for rigging the equipment], it was my last chance to make any changes, I said to the director in the rehearsal room one lunch break, 'Just tell me nine different words to describe each of those nine plays', and on one of the plays I got 'Victorian opera', and this was new to me. She had done four weeks of rehearsal, she now knew her tune and how they [the plays] ran together as well.
>
> Okay, so everything I put on [the lighting plan] for that play is rubbish but that's fine, we'll just go away and work something else out. I now knew exactly what the plays needed from her one-word descriptions. I just had to action it and action it quickly.
>
> It was quite interesting how different it was from her initial, 'This play is about this, this play is about that'. Now it was like, 'No, this is that way, and then we go to this place and then we go to this place and then we go to this place'.
>
> That's all I needed was one word, and I still think that what we got on stage today is relevant to those nine words. Well eight actually. When it came to *Shadow Play*, she said, 'Don't know, let's make it up when we get there'.

Learning from collaboration

Several of those interviewed were keen to talk about what they had learned from their creative collaborations. Trust aids collaboration enormously.

The UK-based physical theatre company Frantic Assembly[1] has provided a trusting environment for Natasha Chivers to make some striking and unconventional lighting designs:

> **Moran:** There's an ambition to use light really powerfully [within Frantic Assembly].
>
> **Natasha Chivers:** Oh yes – it is amazing. I mean some of the best ideas and some of the best shows I've done are theirs, and you learn from it. They are both really good – Scott [Graham] used to light the shows himself, but they both [Graham and Steven Hoggett] love light. And when I started there was lots of really strong stuff going on, with floor cans and really strong diagonals, and really playing with light, all sorts of stuff going on. Incredibly bold.
>
> And I did a show with him about a week ago, – and it's another one – [I have] a big idea that might not work, trust me. He had the idea that the set would be covered in anglepoise lamps. I then had an idea that the anglepoise lamps would light the show. And then [after some R&D] we adapted them. I was a bit nervous. But we lit the show with a bit of backlight and the 18 anglepoises and a few specials from the grid. And there was a point in the show when one of the actors turns the anglepoises down – most of the anglepoises are down the sides and all quite precisely focused – so now we have all the key light for the show pointing away from the stage – and then the performer is going to turn them back up and hopefully they will be pointing in the right place. I mean this is the largest part of my rig – so I was a bit nervous about it. And she was very good and it worked, and it looks fantastic – I'm really pleased with it.
>
> So what Steven [Hoggett] has is that winning combination of great ideas and practicality – and that's fantastic. I mean we've all worked with directors who say 'this is what I want to happen and I don't care how it's done'. Instead Steven goes: 'that's great, now how can we achieve it?' And to work with someone who is fully engaged with how those great ideas will work is fantastic. Steven gets that whole picture. And what they [Frantic Assembly] do is they create work that cries out to be lit, but also if you see an opportunity to use light, they want you to use it. And if there is a tweak that can make the piece work better with light – say if it's about moving to a different diagonal – they will always respond to that.

So far the collaborations we have focused on have been with the director, but the relationship with the set designer can be just as important. Several of those interviewed here have long-standing working relationships with particular set designers: Paule Constable with Rea Smith and Peter Mumford with Hildegard Bechtler, for example. In this section on learning from collaboration, James Farncombe talks about learning to give himself time to see, from a different tradition:

> **James Farncombe:** I had a really interesting experience with a German stage designer Johannes Schütz when we did *The Three Sisters*[2] at the Young Vic a couple of years ago, and he built into his sets a massive lighting fixture. [Because of this, lighting the show] was a collaboration. I couldn't claim to be the sole LD on that show.
>
> He did this incredible thing where he kept saying to me, 'I'm just going to sit and I'm going to look, sit and look' and he'd say, 'I think it's a bit dark. Up 2%.' So we put this massive light box across the whole of the thrust stage and it was just packed full of about 180 fluorescents – producing a sort of massive, diffuse, cloudy light over the whole stage. And we'd put it up 2% and we sat and looked, for 20 minutes, and then he'd say either 'Mr James this light is very good' or he'd say, 'It's too bright, stick it back down again.' Or, 'a bit brighter.' And we'd sit and we'd look and it was magic because the whole process was slowed down.
>
> We sat and we looked and we saw a few actors moving around from a few different seats in the auditorium and it was the first time that I just had time. He insisted that we had time – to just sit and absorb it. That was a massive lesson to me because I think there is an impetus to be as quick as you possibly can. That comes from somewhere else. That's to do with producers and schedules and all that sort of stuff. That's not the concern of an artist, is it? An artist shouldn't have to be bothered about how long it takes to make something.

The theme of time to develop and really finish the work will return, as will concerns with the conditions required to make art. Farncombe, however, went on to echo Constable and others on the importance of early involvement:

> **Moran:** At what point do you get involved ideally?
>
> **James Farncombe:** I'm thoroughly intimidated by the blank sheet of paper; I think the white card model is great, that's a very good point (for me to get involved).
>
> But the few times I've been involved in the conversation right from the word go, it's been so exciting because suddenly all bets are off, you can do

anything. That's the privilege of the director and designer I suppose. I'm working with Sean Holmes and Jon Bausor, we're going to be doing *Bugsy Malone*[3] at the Lyric (Hammersmith) next year (2015), and we went away for a weekend to Sean's place in the country and that's the first time for a long time that I've been in on a project at that early stage. No set, we're just talking about how we're going to do it. And we talked about anything and everything. Fifty percent of the conversations came to nothing, we'd just go off on huge tangents, but that's the sort of collaborative blue-sky thinking that we as LDs, and especially those of us who are rooted in the technical background, have never been privy to before. So that's really brilliant and I hope ... It will be interesting to see what the end product is like, if it's any more harmonious, if the composition is any stronger because of that.

A creative place in the mind

A recurring theme for some of the interviewees was the need to find a creative space in your own head in order to make the work. Mark Jonathan combines his career as an international lighting designer with being a qualified ski instructor. He uses his sports training to get himself into a 'special [creative] place' for the lighting session, even when conditions are not ideal:

Mark Jonathan: When it's working really well you find yourself going to a special place in your head. I talked to a younger friend of mine who's a great creator, Jon Driscoll [award-winning video designer whose professional career started in lighting design], years ago about going to this place. He knew what I meant. That tingling excitement you get when you start to play with ideas about ways you might do something.

The same thing should happen at the lighting session or even at the rehearsal; that you're able to go to this creative place in your head despite the fact that there are distractions. People are jabbering at you in one ear and there's somebody saying something in the other ear, you're up against the clock and your programmer may not be quick enough to keep up. You still have to try and be in this creative place where you know what came before and what's coming next and that you can think through the piece while we deal with this moment. I love that. You're tingling...

I'm talking about going to a place, in my head, you've got to open the door to go into that creative place. Of course, it's great. When you look at the greatest creators, they had to work to schedules. Haydn would be given a week to write an opera or do something for a religious festival and had to keep to the timetable.

We're up against that, we have to be creative against a timetable.

Moran: What helps you to get to that place?

Mark Jonathan: Well, you've done the preparation, you've learnt the piece, and you've done your prep [plans, focus notes, cue synopsis and other lighting paperwork]. We're always up against it, I talk about getting off planes and rushing to a rehearsal. You must never let those things compromise the work – that you're tired or overworked – if you've done the preparation. I have to be really careful myself that I don't have a self-fulfilling prophecy of negativity where I go, 'I know I'm going to screw this up. I know I'm not going to get it right.' That is the first step to getting it wrong. You've got to believe in yourself without being arrogant about it. Woe betide you if you haven't done the prep; that's when it goes wrong.

Moran: Does any of that come from your sports training?

Mark Jonathan: Yes, totally.

Moran: The visualisation thing?

Mark Jonathan: Completely. You can visualise negatively as well as positively. I know that; absolutely. My sports training and my lighting design practice cross-fertilise each other. ... You can improve your skiing performance in this room and go back out there and do it better but you can also destroy it if you think negatively. I know I have to work in that same way with lighting design. Having worked with so many different rigs and stage designs I find it very helpful that I can think in 3D.

Natasha Chivers recalls being in that special creative zone whilst re-making *Macbeth* in New York:

Natasha Chivers: I caught myself – the way I was sitting – in New York recently, when we were there making *Macbeth*. You know, when you get to that point when you know the rig and you don't need the numbers. And I was sitting right back, and just constantly talking – when you're in the zone and have a programmer who can keep up you can do that – and you don't necessarily know where you're going but you are confident that it will be a good place.

In the middle of my interview with Jon Clark, when we had been talking about how to counter the argument that good lighting should go unnoticed by the audience, Clark began to talk about light as one voice among many in a theatre piece. His contention is that when it – the creative process of making a production on stage – is really working, it's like poetry:

> **Jon Clark:** It [the lighting design] is just a voice in the piece of theatre you're making and in that sense I suppose it is poetic. It is not trying to stand out to be descriptive but it's melodic I suppose – it has a rhythm and a melody in the same way that poetry does, and a pace and it's conscious about its rhythm, that the language and the rhythm of the spoken word sit together rather than the prose being measured and descriptive.
>
> **Moran:** I think that's a good analogy. Natasha [Chivers] started talking about the same kind of ideas using [the analogy of] improvising music, so that when it's really cooking, not just the making of light on stage but the making of the production, the lead is passing around different people within that group and sometimes the lead is lighting and sometimes it's somewhere else and sometimes lighting is providing the same sort of thing as a drummer, who's just sitting there holding the rhythm, and sometimes it's playing the lead line.
>
> **Jon Clark:** I think that's a good analogy. It's just one element of a bigger thing that you're trying to make and that you're making a piece of theatre not making a piece of lighting, that's the distinction. So in that context invariably there will be moments where the lighting can lead but it's an ebb and a flow and the baton is being passed around all the time.
>
> I suppose that analogy of improvising, that's when it's most rewarding I think, when you're making it. It's when you're in that moment and you know exactly what you're doing but you're responding to what you're seeing on stage during a tech or a preview or whatever and you're assured of where you are and what you're doing and you can just play and craft and improvise. ... To end up where you are ... but it's not necessarily where you thought you might be, but you're going with the flow and what you end up with is something, when you play it back later, you're not quite sure how you ended up there, but you own it. That is poetic I suppose.

Making special moments: Punctum

Clearly some images on stage have a greater impact on the audience than others. Despite the ephemeral nature of live performance, some stage images are retained and recalled by audiences for many years. More often than not, light has made a significant contribution to those images, but it can be hard to describe how that has been achieved technically or aesthetically – even using the technical language of theatre.

French philosopher Roland Barthes, writing about photographs in the context of photo-journalism and advertising in his short book *Camera Lucida*, used the notion of *stadium* and *punctum*. Barthes was writing about the analysis of the static image, while what we are usually talking about in this context is far from static. Even so, several of the lighting designers interviewed were attracted to his notion of *punctum* – the thing that jumps from the page/stage and almost literally 'pricks the perception' of the viewer/audience member. Barthes' punctum escapes language, and communicates with the viewer/audience on an emotional level. Barthes wrote: 'it shoots out of the [image] like an arrow and pierces me' (Barthes, 1980). For many reasons Barthes' analysis is not in any way able to provide a complete understanding of the image in performance; however, *punctum* is a useful shorthand for that thing in a stage image that makes it stick in the mind of the audience member, and this is how I have used it here, first with Hugh Vanstone:

> **Moran:** Moving back to the art thing. Roland Barthes was a French philosopher writing about photos and photo-journalism. He wrote about a technical quality of image which a good photo journalist or editor will just pick out of a range of images, if you like a technical correctness of the image to do what it's supposed to do, but then in a few very special cases there's something within the image that jumps out at you and makes you go, 'yeah that's the one'. Are there particular cues that you get that kind of feeling about?
>
> **Hugh Vanstone:** Yes I think there are absolutely. The terrifying thing is, I've got to admit to you that there are many times in my own work when I think I've achieved that, and you go 'yes that's it', that moment you were talking about earlier on when you go 'yes that it, that's right, that's right'. ... Sometimes I don't even know how it's done, it just happens!
>
> Thinking about photography, what does make a remarkable photograph or an exceptional photograph? It's usually the subject, particularly if it's a human, just something indefinable about a look, or if we're taking it back into theatre, performers. Performers never cease to amaze me and that whatever it is, that extra 10%/5% of magic. Does it come from the eyes or the brain or whatever? That can communicate that thing with the people in the house. ... there are times it lifts a scene or a song that's good but

nonetheless pedestrian, into the extraordinary – hairs on the back of your neck moment. If you could bottle it, it would be great!

James Farncombe talks about the same kind of tingle moment here:

James Farncombe: There's a sort of 'ping' moment when you're creating lighting where it comes together and my analogy I suppose, which is a little rough round the edges, but it's a bit like taking a bare piece of wood and you put primer on it and sand it back to remove some imperfections, sand it a bit more so it's a little bit better and take a few more bits away and eventually there comes a point when there's a really nice sheen to it, a luminosity to it. It's really hard to know how to do that 'cos it's slightly different, the levels or the colour is slightly different, for every show and every set. I know when I see it. Very often it'll happen and I'll see it the course of a preview and I'll identify the scene that I think is working and we'll go back and set all the [appropriate] levels in other scenes to match that one scene.

This echo's Johanna Town's point about discovering how to light the piece in act 2, and then having to find time to go back to act 1 to make that work too. Neil Austin talks about recognising the potential for these special moments, and then working towards making them, creating the 'ping' Farncombe talks about:

Moran: There are moments when it really works ... Are you surprised when that happens? Do you plan for those?

Neil Austin: I think it comes back to your earlier question, do you know exactly what it's going to look like in advance? No. Some of the best moments have always been happy accidents. Some of the most dynamic have been happy accidents. The skill is recognising them and taking them – grabbing them.

There's always that moment when you ask for a channel and someone presses the wrong button, the wrong number, and you go, 'Hang on a minute. That's interesting. Let's do that version instead.'

Sometimes they're planned, sometimes to within an inch of their life and other times they are discovered along the way and you say, 'I could do a bit more of that,' or 'That's an interesting idea but that's not the way to do it properly but I could take the idea and [develop it more effectively].'

Austin goes on to remind us that making theatre is very often the art of the possible – and that knowing what is possible is another skill that comes with experience:

> **Neil Austin:** That's where the skill comes in – and understanding. The experience of having rigged the lights yourself is the difference between someone going, 'I've got this amazing idea. Why don't we hang ten 5kWs up in that corner over there? It'll look brilliant', when there's no rigging there, there's no power, we don't have ten 5kWs and we won't get them for three days so the idea won't go into the show for another four days. Or having that idea and going, 'this one thing here would actually be just as good, and we can get that done in the half an hour we have, and it will be in the show this afternoon, and we can see if it works.' That's part of the skill as well. You don't know everything in advance. It's a process like any other where there are discoveries on the way.

Personally I don't think you have to have rigged lights yourself to gain that skill, but you do have to have had active experience of making things happen quickly on stage.

Paule Constable's take on the idea of *punctum* was thoughtful as ever:

> **Moran:** There's an idea from Roland Barthes – looking at press photos, you know, Magnum-type press photos, all technically very good, looking for the thing that is different in the ones that stand out. He calls it *punctum*. It seems to me that that might be a useful thing for us to think about.
>
> **Paule Constable:** Oh yes! Yes I think there is something key there.
>
> **Moran:** I think I see it as noticing the thing that pops out of the stage picture whether you've consciously created it or not...
>
> **Paule Constable:** And then what's interesting even if you haven't consciously created it, is going: 'OK why does that work?' That's what's so brilliant about what we do. There is always that massive element of surprise, however many years you've been doing it about how something actually sits in the space when it's the real thing. How it responds to light – however much you've worked with painters, however much you've worked with costume, the real thing can always side-swipe you.
>
> You know, you have all these ideas and you put it all on stage and go: 'whooooops!' But then it's amazing how quickly you can go: 'just cool that down, just bring that up' or 'that angle's too steep. It needs to be this.' And

> then something that was skewing you and making you swear, becomes something else.

Implicit in this notion is the technical and aesthetic understanding that comes from a great deal of experience, and builds the confidence that you know how to get from something 'that was making you swear' to 'something else.' Constable gives some clues as to where some of that came from for her:

> **Moran:** And how important is it to you to build on the accident – to see its potential?
>
> **Paule Constable:** I was a production electrician for quite a lot of people, and I know for instance Rick [Fisher] – his shows were about possibilities, and he'd put stuff up and see where it led him – and I know that's not how I think. But sometimes if you are in a mess, again, it's like taking all the lights away and let's do something surprising with this.
>
> **Paule Constable:** In order for that special thing to happen, I do quite often at a block [cue – technically a point where the lighting designer can be sure that doing something radical will not affect the look of the rest of the show], I'll go for one light at full, and sometimes that might just be an arbitrary mad choice, just so you don't go to black-out. So I might go something like: 'what would that downstage left [moving light] look like at top in a vile green?' and sometimes it's great!

Constable is ready to try the oddest thing – any light is better than no light as a starting point, while Austin makes the point that you also need to know what you can do with what you have to hand – in the time you have to do something in.

A lighting designer may not always know how she or he got to a great place in their process, but the key is to recognise that you are there and to build on it. As a theatre lighting designer, you have to be able to see potentials, both in what you have planned and in what you discover. To aim for and to recognise when your work communicates 'beyond language' – while still functioning as one of many voices in the piece of theatre you are making (to paraphrase Jon Clark) or playing in the right harmony and rhythm, to paraphrase Natasha Chivers. This is one of the markers of Bryon's *active aesthetic*. As she writes in *Integrative Performance*:

> Choices made are both creative and technical. The best choices are not complicated: however there is a certain level of complexity involved in the planning (Bryon, 2014, p. 193).

Bryon writes about movement performers, but her words seem particularly appropriate here to describe the active practice of lighting designers too.

Precision

Another of the markers of lighting design with an active aesthetic is a willingness to take risks. One way in which this is often made manifest is precision – the placement of light and darkness on stage, the timing of cues to work with other elements of the production – sound, automated set-changes, magic tricks, music and the voices and bodies of the performers. Technology has helped lighting designers contemplate working in this way – much of the older equipment used by previous generations would not have been able to achieve the repeatable precision of today's equipment in a number of areas. More than any other part of creating a lighting design, this is an area that requires close collaboration, between the lighting designer and the director, but also with other members of the creative team and of course the performers.

> **Moran:** A lot of your work seems to me, from the outside, to have a precision to it – by which I mean that very often for a particular cue to work the performer has to be in the right place. Would you agree?
>
> **Natasha Chivers:** Yes.
>
> **Moran:** How do you go about negotiating that?
>
> **Natasha Chivers:** I've slowly got better at working in a strategic way and I have learned to care less about things that I can't control. I've also learned to be a bit more cunning about how I get the things I do want to happen. So I might start by saying something like: 'how precise do we want to be?' I'm a bit manipulative and might dangle carrot like 'we might have the opportunity to make some really striking images here if we can be precise, get the actors to stand on marks if that's appropriate. You'll need to work with me on that.'
>
> And it's good to say, 'well it may not be appropriate, you may not want [the performer] to lose that freedom' [lost when the performer is constrained to be in *the* right place for the lighting]. Then they often go, 'well what do you mean? What do I get if I do this and what do I get if I do that?' So you show them, and say, 'but it might not be what you want' – you know, and I sometimes pretend that I don't really care. I've found that if you pretend not to care then people will start caring for you, because they

> then worry about how to make you care more. Terrible admission but it seems to work.

In chapter 2 Peter Mumford spoke about how he often aims for a filmic look, or at least a look inspired by film. In order to create those looks – for camera or for a live audience – very often it is necessary for the performer to 'hit their marks'; in other words to be in the right place looking in the right direction.

> **Moran:** [What you have been talking about] would tend to require a degree of precision, and therefore for the performers to be in the 'right' places...
>
> **Peter Mumford:** Yes. I find with most shows you can do that. I like also the fact that these days you light over the technical process. ... I love working with actors when you're lighting it because you can actually give them light. I mean I always say, [as an actor] you should know where the light is, like you know where the chair is, and if you're going to go into shadow then know that you're going to go into shadow. Which doesn't mean you don't try and work with them...
>
> There are some cases where the director has decided to allow free blocking, and you've got an actor who wants to work like that. So doing *A Doll's House*[4] with Janet McTeer and talking to her, we talked about having it really like moths around the flame: well that [didn't work] – she was bouncing off the walls, and there was just no way – and you just had to open it out and let her do what she wanted, because it was different every night.
>
> But for the most part, actors like to have a precise track through a piece, they like to know where the other actors are going to be, and the kinds of nuances they'll put on it will be largely internal, it won't be about, 'ok, now I'm actually going to do this from the other side of the room.' So you can then work with them, I think, and teach them to love the light.
>
> Also, if they know they are going to look good in that light then they'll get in it. But there are some ... Michael Bryant, you could put a slit of light ... and he'd be in it. You wouldn't have to say anything. It might not even be meant for him, but he'd find it and be in it. David Suchet is very good too. He knows exactly where the best light is, he does really know. You will never see David Suchet in a dark corner unless he means to be.

Thirty years ago it was rare for a lighting designer to talk to an actor about getting into their light. Not so any more. Besides Chivers and Mumford,

Austin, Farncombe and Clark all talked about the need to discuss with the director and with the actors the whole business of playing with their light.

For some theatre makers and performers, the kind of precision that requires the performer to be in the same place every night works against other things they find more important, such as the notion of each live performance being a unique event. Hugh Vanstone spoke about this apparent contradiction in the work he has done with his long-time collaborator, director Matthew Warchus:

> **Moran:** This idea that for the show to work most of it has to be the same every night? Do you find that is restricting?
>
> **Hugh Vanstone:** Yes and no. One director I've worked with a lot, Matthew Warchus, is particularly attuned to that issue. The very first thing I did with him was *Peter Pan* at the West Yorkshire Playhouse years and years ago. I'd never met him before, and virtually on our first meeting he said, 'oh by the way I don't believe in blocking, so we're not going to have any'. And I said, 'how are they going to get out of the door then? If they're standing over there when they're about to go out of the door, they're not going to cut a hole in the wall are they?' We had a good laugh about that and so we've gone on in life and of course he does believe in blocking.
>
> As he's gone on and become a more experienced director what this has become is that there will be bits that we call 'free-form', or whatever his word of the day is, which means, 'within certain constraints in this scene I want the actors to be able to roam freely within this triangle, or they can go further but we're totally expecting that if they do they're going to be dark'. And of course the actors love it and it's good to keep it all alive and he will deliberately put in free-form sections so that things can breathe and change and there are other things that are absolutely rigid, 'you must be standing over here before the 10-ton wall flies in and crushes you!'. It's obvious
>
> The latest example with massive freeform was *La Bête*[5] at the Comedy and that was very, very unblocked. Partly to cope with that we tried using Autopilot [a preparatory tracking system that enables automated fixtures to 'track' a moving target, which can be a performer]. That worked with varying degrees of success, not massively actually, but it sort of helped a bit. I quite enjoyed the challenge. It's liberating sometimes to work in a different way, and so [we decided], 'Ok we're going to make something that's architecturally pleasing, that's about the right brightness and mood for this scene, and then over to you Matthew, because that what's it is, and

if you let them go in a dark corner or if they want to go in a dark corner, more fool them!' And of course then what happens is they all find out the good sweet spots on the stage, acoustically and for light and Mark Rylance is like, 'ah! Found it!'

Isn't that extraordinary though that some [actors] have a great sense for it and some of them have absolutely not a clue and don't even know if they're in it or not, let alone find it! It must be something to do with photosensitivity!

Johanna Town works predominantly in drama and – as she says herself – usually sees it as part of her role to find a solution that puts the right light onto the right part of each performer wherever that actor choses to stand. But sometimes she works in a different way:

Johanna Town: I just did a show this year [2014] in fact that was set in an Israeli prison, Palestinians in an Israeli prison, [*The Keepers of Infinite Space*[6]] and that was all lit by practicals. We did enhance it, but we never really wanted the audience to see that it was enhanced with real theatre lights. …

It was really enjoyable to engage more with the actor. I had to explain to the actors, you do have to find your point to stand, 'it has to be two inches further upstage', otherwise that lightbulb is never going to light your face and I need you to look up into that lightbulb, it enhanced the uncomfortable torture aspect of the play; everybody, actors and director, was with me on it.

Moran: So it's part of the project right from day one?

Johanna Town: Yes, it was part of the project. I'm really not good on manipulating actors into place; I'd much rather go, 'I'll solve that.' You see it with a lot of actors, they go, 'should I be here?' You go, 'no, stand where you want to stand, that's my problem, I can move that. You're looking all unnatural standing in that place, because I've put that light in the wrong place. Go back to where you feel you should be.' They all seem a lot more obliging.

But in the case of *Keepers* I had to be quite strict and we had to keep stopping the tech every five minutes going, 'This scene, you've got to hit that mark, floorboard there, and nail this, and you've got to hit that floorboard there', in order to shape them and make it look as if we really were lighting from the practicals and the torches and everything else that we were wheeling around lighting wise, but then they enjoyed it because they felt that they were being tortured under a lightbulb! So they were all up for

> making sure that it worked because it helped their character of course in the end.

Town here changing her standard practice to suit the aesthetic and practical demands of a particular production, and being excited by doing so, is a good example of the flexibility of approach of most lighting designers – the chameleon role Mark Jonathan refers to.

Precision includes the fear of falling

Along with taking risks, though, comes the fear of failing. Nobody I interviewed was quite as eloquent about this as Jon Clark here:

> **Jon Clark:** It's that Svoboda quote about, to paraphrase massively, but sitting in an empty theatre looking at a space and having the fear that he won't be able to crack it the next time and that kind of search for creative [inspiration]. It comforts me every week of the year that someone of that kind of genius is gripped by that fear that everybody knows, it's kind of alright. But it's not, but that's the pay-off, it's the Faustian pact of what you get at the end if you get it right. That the fear that this time you won't crack it, that sitting at the end watching the Press Night or worse a couple of previews before Press Night, when you know, not that it's a terrible piece of work, but that something hasn't clicked. Maybe it's not you but I always find that impossible, it's always me in my head – however much you analyse it! There is sometimes an overwhelming disappointment that it's just not what you wanted it to be.

Technical collaborators, programmers, stage management and others

So far in this chapter we have discussed collaborations with other members of the creative team and performers. The other, equally important group of collaborators includes lighting crew and follow-spot operators and the DSM or deputy stage manager, who is responsible for calling each cue of the show.

We heard in chapter 3 how Pyant assembles a team of lighting people to ensure the machine of his lighting rig is ready in time. Here he talks about the importance of people he does not choose himself, stage managers and follow-spot operators. These are people who, when they are good at their job, make precision in the placement and timing of cues possible.

Paul Pyant: So you've got inspirational stage management. My job becomes so easy if the DSM is on the ball. I mean not just on the ball, but ahead of you, which I think is terrific. I met recently two or three who are. The team on *Charlie and the Chocolate Factory*, just, I mean, wow. It's extraordinary what they can do. And what's more they do it night after night. I leave it and I go back and see the show [after months] and you still see that [very complex] show being churned out with a degree of accuracy that is just amazing. It's breath-taking.

The other people I have the most enormous respect for are the follow-spot operators because [at Drury Lane] they're next door, they're half a mile away[7], and their accuracy is just stunning. Bearing in mind I was the worst follow-spot operator in the world. I admire that, I really do. And it is about that relationship, that these people all know their job and they're phenomenal.

Pyant also mentions how important it is that the Lx crew look after the installation, particularly if it's based round moving lights, picking out for particular praise Steve McAndrew, the chief electrician at the Theatre Royal Drury Lane, who leads a dedicated team that look after his rig (and other things) on *Charlie and the Chocolate Factory.*

Except in the smallest venues, no lighting designer realises their work without technical assistance. In his 2012 PhD thesis Dr Nick Hunt argued for, amongst other things, a broader acknowledgment of the virtuosity of technical operators working in theatre. It was during informal discussions a few years before the publication of this thesis that I first came across the designation 'technical artist'. Hunt's context is the role of the light board operator 'playing' the lighting, and he references early light boards that were built around theatre organ consoles, where the operator literally played keys to make lighting changes. As we will shortly see, this is not a way of working that lighting designers here are particularly interested in; however, the term technical artist has caught on and it is now rapidly gaining currency in UK theatre. It speaks to the skills, knowledge and aesthetic understanding of people often referred to as techies, sometimes in a rather derogatory way. These are people who have perhaps for too long been taken for granted by some in theatre and beyond. Here is Katharine Williams talking about the lighting crew who work behind the lanterns when she is focusing her rig:

Katharine Williams: [I]t's that lovely partnership in focusing when you have someone who is good, cares about lighting and is interested, then

> you get to explain what No. 1 does, give a bit of help with the next one, then just guide through the rest. And for me that's the best – it's the right way anyway. It's trusting the technician to be the creative that they are. We are all technical artists.

Hugh Vanstone is also generous in acknowledging that his work is not a solo effort:

> **Hugh Vanstone:** I love being helped as well, that's my other discovery of going through life – maybe I'm just working with better people now – but I love it when I've got a great programmer and a great associate and all I have to do is say 'make it more choppy' and 'this bit's got to go like this,' and you come in tomorrow morning and it's done. And they do a brilliant job. I love it.
>
> **Moran:** And they've got your style.
>
> **Hugh Vanstone:** Yeah sometimes. Other times they completely bring something to it themselves.

The role of lighting programmer is increasingly important in UK theatre. Bruno Poet is not alone in asserting that programmers in the UK are the best he ever works with and make it easier and more enjoyable to do his job well. Mark Henderson comes from an older generation, but that does not mean he values his collaboration with the programmer any the less:

> **Moran:** What do you think is the role of the programmer in your work today?
>
> **Mark Henderson:** If it's a musical – then they are key. ... They are so vital for their speed and efficiency, but also for their input as well. They are generally very experienced, and they've seen stuff and they know what the lights can do – probably better than I do. They work with them all the time. In that environment they are able to offer up suggestions and have an input, which is great.
>
> But you do work in a lot of different circumstances – As you know I'm on a job at the Royal Opera House now and the rig has, I don't know – around 200 moving lights. You're working with a programmer who works with that rig every day. It's the same at the National (Theatre). The good ones really know their rig and how to help you get the best from it.

Moran: So you rely on the programmer.

Mark Henderson: Yes – yes.

Here are first Lucy Carter then Paule Constable on why they need a good programmer for some types of show:

Lucy Carter: I think now that a good programmer makes a good design for me. I use quite a lot of detailed effects, not in terms of traditional effects, but kind of weird things. I couldn't programme a desk any more so I need somebody who knows how to do those things for me. It's more important to me than having an assistant, and they kind of do become your assistant because they've got a creative role too. You can say 'I want this,' not in terms of the look of the stage, but in terms of timings of cues and effects. 'How can I get that cue fade to look like it jumps in rather than fades smoothly over 15 seconds?' [That's part of] a creative role for a lighting programmer.

Paule Constable: Good programmers are great. Absolutely brilliant. ... I think on a big show you need a good programmer, because it's too much. And it's more than data input. It is a collaborative process. When I do the big musicals I have a team who I pick very carefully and we work together and we make something together, and I think it's a really healthy way to approach work – to take people on a journey with you. You can have all the ideas in the world, but if you can't take people on that journey and share those ideas and get them realised, what's the point? So actually I think it's a better thing to empower people.

I am quite hands on, and I don't just say, 'just make 15 chases, I'm going to go away and see you next week.' I don't work like that. I make structures and parameters and things – it's just my way of doing it.

Moran: And [your programmers] don't get in the way [of the creative process] for you?

Paule Constable: No. Only if they are not very good and they're a bit slow, and then ... you have to find a new rhythm of working. Which is hard if you have a producer or a director who wants things like this (clicks fingers). But you have to cut your suit to your cloth don't you?

Ben Ormerod has a very clearly defined relationship with his programmers – one that is quite different from that which Hugh Vanstone talks about.

In this conversation we are comparing the roles of lighting programmer, working alongside the lighting designer in the tech, and the re-lighter, who is responsible for reproducing the lighting designer's work in the absence of the lighting designer – for example if the production tours or is part of a repertory season:

> **Moran:** Could we talk about interpretation, something that is required by a good programmer, or by a re-lighter reproducing your lighting design in another venue?
>
> **Ben Ormerod:** Yes to the second, no to the first. I work a lot with Andi Davis as my programmer and he is meticulous in achieving exactly what it is that I want, without in any way interpreting what it is that I want. Even though I will sometimes – because we've worked together for years and years and years – so even though I will say, 'Find me something up in that corner that's available to do a three quarter on that boy over there', that's not the same as re-lighting when I'm not there. It is hugely different.
>
> The relationship that I look for in my programmer is not one of an interpretive musician. I know it is for some lighting designers, but it isn't with me. I know exactly what I want. But, I find that if I'm working with a programmer who understands me, then when you hear us work together, you might think that I'm having the *Beleuchtungsmeister*/lighting designer relationship with the programmer, but I'm not, it's just that they know what I mean.

The locked down show

As we have already noted, in the UK and North America, current practice for almost all theatre shows of any size is for all the lighting cues to be played back on a lighting computer (console) at the push of a button. With a good programmer, even the most complex multi-part lighting transitions can be accomplished this way. This means that in theory at least, every transition between one cue state and the next will be accomplished in exactly the same way every night. This situation is frequently referred to as the 'locked down' show. As already mentioned, for some the idea of 'liveness' is to some extent at odds with the idea that any element of a show should be the same every night, yet most theatre lighting designers expect their lighting to be locked down. But that does not make it quite as inflexible as it might first appear:

> **Moran:** Do you think it affects the way you light the show, the fact that you know it is going to be locked down?

David Howe: Yes. It gives me a structure. I can understand Paul Anderson, when he used to do [design and operate lighting for the theatre company] Complicite and every performance was different, and they would move things around constantly, and the operators would take their own cues. I completely get that.

But for most productions, if you walk away from it, as in you've done your job and you're on to the next, then you look back on it and go, 'I'm leaving that as I would like to see it when I go and see it again in x months' time'.

But if the show drifts, or the show changes, then I move things.

Intention and flexibility

How does that happen? What happens when, as the 'show drifts or ... changes', a lighting transition no longer fulfils the original intention? Is the solution to move where it is cued, to change its timing, or to tell the actor to go back to doing what he was originally directed to do? The lighting designer will not be on hand to help make that decision (or to notice that the transition no longer fulfils the original intention for that matter). So who makes that call and how does the lighting designer ensure the integrity of their work each night?

David Howe: If I'm told that we need to move a cue that's fine. We move a cue. I don't argue about it. [If I get a call from the DSM]: 'The director was in the other night. Do you mind if we move cue X to this point?'

'No, just change the timing on it and make it three seconds rather than seven. That will work for you.' I have no problem with that at all.

Moran: Do you leave the SMs with freedom or not? How do you deal with, for instance, the whole thing of where this cue actually goes, whether it goes on a particular word, in a particular line, or a movement, or something else?

David Howe: Sometimes you say, 'Put it on the words, "To be or not to be". Put it on the word "Not"'. Good DSMs are smart. They come back and say, 'in the tech you put it on "Not" but they actually move on "To"'. 'Well, they used to move on "Not", because that's where it says I've got the move here', or the underscore starts there, or the sound effect starts there.

Moran: Because a lot of the time these days we are going with actors' movements, and so is the underscore, so do you let the DSM know that it's with a movement?

David Howe: Often, yes. If it's that reliant on it, yes. If then she's got four cues that happen, 'To be or not to be', or whatever it is, if there's cue, cue, cue, cue of various different sorts of course it's silly. 'Put it all on the one point and I will put a delay on. Sound, will you put a delay on yours?' 'Yes, it's fine.'

Yes, I do say [to the DSM] what it's about. I do tell them what the cue is for. I don't say, 'Cue four goes on this word, cue six goes on this word'. I go, 'It's the moment when they cross downstage, which I think is at this point'. And they go, 'Is it?' I go, 'That's what they did in the run-through'.

Moran: Expanding on from that, do you give them what the cue is doing?

David Howe: Yes. Maybe, 'It changes the backlight from this colour to that'. 'It's a change of this or it's a change of that.' 'It's pulling the focus down over two minutes, to this moment where they do that scene downstage on the couch.' Oh, yes.

Moran: So it is about including them in the process?

David Howe: Well, they're giving a performance, in the same way the actors do, and the same with the follow-spots. My speech to the follow-spot operators is always the same. They're the last people into a rehearsal process, also the first people we shout at when it's not absolutely right and subtle. But, if they get it right, they're absolutely giving a show as well, the same way the DSM gives a show. There's a little breath moment between the end of a song, or the end of an aria, until the next recitative or something, and if a cue goes on the breath – you can't define it. It's not one and a half seconds after this. It's not on the upbeat. It's not on the downbeat. It's somewhere in-between. And if they get it, they're giving a performance. So yes, absolutely, I think they're not just automatons.

This idea that the technical operators are giving a performance is part of what is behind Katharine Williams using the designation technical artist for everyone working in a so-called backstage role on a show. On some productions this may seem to be overstating responsibilities; on others it definitely is not:

Jon Clark: On [the short-lived West End Musical] *I can't Sing*, the DSM was phenomenal. There were two numbers in it which should be [cued automatically, using] time-code. But she was absolutely vehement, no time-coding. She was going to call it and she was good and so fearless about it so I said 'OK, you prove it', and she bloody well did! She was very musical and really disciplined and she did it, and her reasons for [those two numbers] not being time-coded were really honourable. 'This is my craft and this is what I do, I don't want my job to be taken over by a machine.' She could do it.

Had the show run [it closed within two months of opening] I would have kept an eye on it and if she'd moved on [to another show] I had up-to-date time-codes for all of the cues so we could at any point put it in and done it to time-code – but good for her!

Much of Paule Constable's early work as a lighting designer was self-operated, for concerts and for experimental theatre companies including Theatre Du Complicite. Now the scale of almost all her work demands a locked down approach. Here is what she had to say on the subject:

Moran: Moving onto the locked down show – especially as you come from a live concert and devised theatre background [where there is very often an element of improvisation required of the lighting operator, responding to the particularity of each night's show], how comfortable are you with having to design lighting that gets programmed so that it can be the same every night?

Paule Constable: It varies, doesn't it? There is a certain kind of show that, especially if it is touring, will shift – move – grow, and you can't hold them to ransom, it would be wrong to. And then you have to think very carefully about how that works, how it's done when it's on the road.

For something that is running for a long time – like *A Curious Incident* or *War Horse*, I have associates who go and see those shows every six weeks. And then ... if there are problems – [it may be] the blocking has changed, perhaps because an actor has got lazy or the part is being played by an understudy ... so you're constantly trying to check in with why that's happening with shows that are extensively locked down. And then when there is a cast change, we generally do a re-visit.

Hugh Vanstone, like Constable, has a great deal of experience with long-running shows, some of which have multiple productions running at the same time. Here is what he has to say about keeping the show 'live' while maintaining the intention of his lighting design as the show changes from night to night and over longer periods of time:

Moran: What about the business of flexibility of cues and cue timings? Do you sometimes regret that you have to give a particular time with lighting consoles at the moment, that don't allow much flexibility?

Hugh Vanstone: Yes and no. Yes I do, but I find it so easy to build in workarounds for automatic cue timings and I'll very, very often do that. I would say I actively try and think up ways of making operators more rather than less involved. [For example] if you're trying to make some fade time-out musically to a speech that's a minute and a half long, I'll often do it in chunks or construct the cue in sections so that it's an overall arc but there are markers within the arc, that if one's a bit longer or a bit slower there are then second cues that go within it. I find it pretty easy to make it all, but it's rarely as simple as, 'oh I just want the fader to move from full to zero in a minute and a half'; there's all sorts of things happening within it, so when I get the idea, I just break cues up.

I love giving freedom to stage managers and electricians. In a long-running show like *Matilda*, they all have freedom to change timings and adjust things, within common-sense constraints. Stage managers adjust things if the performance of one child does something one way and another does it another way, well just change the way you call. I love it. The same with cue timings, feel free. The lighting is there to serve the show and it's a living thing, it's not a relic. Even down to follow-spot operators, I love them and what they can do and how they can sort of breathe with a performance as well. There are those times when it seems as if it's 10% brighter tonight! I love all that, I like it to be a slightly moving canvas. I think that's really good. And it keeps people involved as well.

Reproduction of the work

The conversation with Vanstone went on to discuss the qualities he requires from those who reproduce his work in his absence – which in turn leads to reflections on whether or not the work is art:

Moran: Which sort of leads us on to the whole business of reproduction, trusting other people to reproduce your work... What are you looking for in someone who's going to reproduce your work on tour?

Hugh Vanstone: Someone who loves the show that they're going to be doing first of all, because if you're cynical about it you'll never be any good at it. Of course someone who's got the attendant abilities, that almost goes without saying, but I think the number one thing is when they love the show and they're enthusiastic about the show. There's no bigger turn-off to me than someone who says: 'urghhhhh, going up to Manchester urghhhh'. Ok probably when you've moved *The Sound of Music* 15 times, then you might not be looking forward to getting on the train to go Manchester! But generally it's that involvement and enthusiasm and a kind of sense of fun. ...

> **Moran:** So, to reproduce the work needs somebody who is investing emotion in it? If it can be done without that, then it wasn't art in the first place?
>
> **Hugh Vanstone:** Yes that's interesting point ... is it art? I just don't know. Is all art entertainment? Is entertainment art? Fundamental questions that I really cannot answer!

Well Vanstone clearly feels that his work can only be adequately reproduced by people who are not only technically competent, but are also investing emotion in its reproduction – so perhaps that is pointing towards his work being art.

Paule Constable readily asserts that lighting design – when it is not falling into the trap of being pedestrian – is art. Her work too is reproduced round the world, and she usually makes sure she is involved:

> **Paule Constable:** One of the things that Cameron [Mackintosh] does very well is that whenever there is a new production of one of his shows going on anywhere in the world, you are required to be there. But I get paid very handsomely for the time I give to Cameron, and I get a royalty, and I think it's the least I can do. And I really respect the fact that he's there too, and that input. We try and do that with *War Horse* as much as we can – though it's getting more and more difficult – but you try to check in with it.

So Constable strives to get the time and resources to make the lighting on each show as good as she can, every place that it happens, for as long as it happens in that place.

Rick Fisher, too, has experience of his shows being reproduced and remade on tour and all over the world:

> **Moran:** Going back to the shows you've done over and over – how do you find a satisfactory way of recording (or communicating) not just the technical aspects [of the show] but the intention of the lighting design? Especially when you can't be there to realise the reproduction?
>
> **Rick Fisher:** It's very hard. ... Not everything is done by numbers. It's just kind of what looks good, and it is very hard to record that. You just have to accept that it's going to be different and you respond to it differently. I've seen other people re-create my work and it's been kind of, 'yes, I guess that sort of feels like mine but it doesn't really feel like mine'. It just becomes theirs. And you have to let it be theirs – you have to let go. It's tough.
>
> **Moran:** Has *An Inspector Calls*[8] felt different with each different person that has re-created it?

Rick Fisher: Actually no. I'd like to think that I would mix it better on the day than maybe somebody else does, but *Inspector* survives pretty well, because it is pretty strong. As long as people are lit it kind of works, because it's not the wrong lights it's ... maybe sometimes there are just too many of them on. Or the lights aren't quite hitting where ... As long as they are not talking in the dark *Inspector* looks OK. Yes, it does look better if there's less light on the floor, but actually as the floors have got better over the years, we've actually allowed a little bit more flare to hit the floor, and that's not a bad thing.

Moran: And what about the shows in the big opera houses? By the time they come back out of store and into the rep you will have done so many more things, and the chances are that if the same lighting supervisor is in charge, they may know (or feel they know) the show better than you.

Rick Fisher: It was weird. They revived the *Wozzeck*[9] I did at the Royal Opera House (in 2003) last year (in 2013). Of course I only found out about it through the publicity as opposed to being told – because they don't want you to come back, they are not going to pay you to come back. You have no rights on the show other than it has still got your name on it. I did go back for a couple of sessions, as much to support the person who was tasked by The House with recreating my lighting design, who had not been around when it was created. And it looked good – it still looked fine, and parts still looked strong. But I don't think it looked the way it would have had we been making it fresh, but it kind of worked. But 'was it really like this?' I kept thinking. 'Was it really like that? I can't believe that.' But that was ten years old...

Fisher's self-questioning is as much about 'did we really decide to do it that way?', as it is about 'is it the way I would have made it today?' Surely this, too, speaks to the creation of lighting design as an art practice. Just as no one steps in the same river twice, so perhaps no artist makes the exact same work twice. Ten years on Fisher has changed as an artist and so his artistic response to the production changes too.

It has to be yours

In this chapter we have looked at the lighting designer as both artistic and technical collaborator. We have also looked at aspects of creativity in the moment: Natasha Chivers, so tuned into what she is creating that 'you don't necessarily know where you're going but you are confident that it will be a good place'. Mark Jonathan's 'special place in the mind' as something you have to find even when conditions are not ideal, or Paule Constable 'flying by the seat of her pants' to turn something 'that was making you swear into something else'.

To close this chapter, here is an insight from Chivers, gained from judging the ALD's annual lighting design competition for students and recent graduates, named for one of the first people to be credited as a professional lighting designer in the UK, Michael Northen:

> **Natasha Chivers:** What you see really clearly is people achieve a thing called a 'lighting design' in two different ways. ... There is what you might call a solid way, they may have followed a pattern that they have been taught, or that they have learned from books, a technique that enables them to achieve light on stage that is even when it needs to be, atmospheric and other things, but you don't get a sense of that person in what they present – or in their lighting design. You feel that they are following a formula. I think they have achieved something – they have done a job well – but the ones I'm more interested in are the ones who have let themselves go, and they've done their research and come up with an idea which goes beyond formulas like a general cover. It is something that has come up from somewhere inside themselves and I think that's probably the route that I follow – the way that I work.
>
> Of course you pick up skills and you develop strategies along the way. There is always the big question: 'if this doesn't work, how are we [as in the whole creative team] going to move on?' ... And I think you will always feel more comfortable not knowing where your pen is going to take you if you're working with people who you trust. Director and designer of course, but the whole team.

5 Dramaturgy: Light Telling Stories

The LD's role as dramaturge – guardian of 'the story'

Lighting design can do a great deal to ensure that things are seen by the audience in an order and with a priority that helps them to understand the story in a particular way. This is what draws me towards examining the dramaturgical role of the lighting designer. There are several different interpretations of dramaturge in the English language, and in UK theatre practice. The one I am taking about, in its most general formulation, is *guardian of the story*. To be clear, this is quite different from any search for the original intention of the author. What is interesting and productive is a focus on and concern for how the story is told – if you like, the big picture. It seems to me that this is Mark Henderson's concern when he returns to the final run-through in the rehearsal room:

> **Mark Henderson:** They [the director and his/her close team] are dealing with the detail, which is quite important. So I'm focusing on the main things that are telling the story. But there are other things going on – that need to go on – that are also helping tell the story.

For many productions, the lighting designer is, as often as not, an infrequent visitor to the rehearsal room during the process. As such she can provide a useful perspective on developments that might not be so obvious to members of the production team who are more immersed in what's going on. The outsider view, too, can perhaps more accurately replicate how an audience might experience the production for the first:

> **Nick Richings:** The lighting designer is a person who comes [back into the rehearsal room] with fresh eyes. So you can ask: 'why is this person standing there?' and you have enough authority within the artistic team to be able to go: 'wait a minute, this doesn't make any sense.' And yes I do absolutely believe that that is true. Because I haven't been to every single

> day of rehearsal, I haven't done hundreds of [costume] fittings – and all the rest of it. I've seen the white card and the finished model, I've read the script and been to first day of rehearsal and met everybody, and then I go away for four weeks. And then I come back, and go: 'oh! It's not quite what I was expecting' and you have a chat. And often you get back: 'oh! Wow yes, that's a very good point.' So it is quite useful not being quite so immersed in the production as the other members of the team, particularly costume and set [designers].
>
> Or on the other hand, sometimes you first of all go to the designer and they say: 'oh I know! It's crap isn't it?' (laughs) which is another problem ... 'I've told him that six or seven times but he's insisting on doing it.'

Even lighting designers who make it their business to visit the rehearsal room on a more regular basis – Johanna Town, Paule Constable and Bruno Poet, for example – tell something like this same story of the benefits of 'fresh eyes.'

Done poorly, lighting design can make even the simplest story impossible to unravel. The assertion that you cannot hear people speak if you cannot see their faces well is clearly a massive generalisation – how would, for example, speech radio work if it were universally true? However, studies into speech recognition and artificial intelligence have confirmed what many stage directors and others have known all along, that clarity and understanding are significantly improved when the listener can clearly see the speaker's face, particularly the upside-down triangle from eyebrows to chin. This understanding is key to the practice of all the lighting designers interviewed here, and why they focus so much of their effort on getting the light right on faces. Failing to do this will – as Mark Henderson put it – lead to you 'getting it in the neck!' Bruno Poet talks here about the importance of getting the light on faces right in the context of ensuring the actors can tell the story:

> **Moran:** How important is it to you, in drama particularly, that the audience can see eyes and faces when actors are talking?
>
> **Bruno Poet:** I do think it is important. I think it is very important, because I think you have to let the actors tell the story. Sometimes you can tell the story by seeing the back of someone's head, and sometimes you can tell the story by seeing them in silhouette or in shadow. But I think you have to choose those moments for a specific reason. I think part of the craft of lighting is being able to get the artistic picture you want, while

> still being able to ensure people are well lit. I think it is very important. I get annoyed when I can't see people.
>
> ... [A]nd also we [as designers] are serving what the performers are doing. And hopefully we are helping the performers communicate and connect to the audience. Certainly helping the audience connect to the performers. So, you do need to see their eyes.

Supporting the actors and 'helping the audience connect to the performers' is not the only way in which lighting design contributes to the story-telling. Whilst good lighting design can guide the attention of the audience, poor lighting design can distract with spectacle, or simply fail to alert the audience to what it is important for them to see at that particular instant. Whilst not all performance is narrative, when it is, lighting can play a significant role in making that narrative clear to the audience. Here is Poet again:

> **Moran:** So you do feel a duty to support the narrative?
>
> **Bruno Poet:** In lighting terms? Absolutely. I think there is no point in any of us being there if we are not telling the story. That's why the audience is here, that's why the actors are here. That's why the director has chosen to do that production – they feel that there is something to say about that story. And sometimes you might want to work against it for a particular reason, but generally it's supporting the story and clarifying the story.

In the opening chapters Peter Mumford and Neil Austin both talked about similarities between the role of theatre lighting designer and the director of photography for film. Ben Ormerod talks here about his approach to directing the attention of the audience, using some of the same notions that come from that analogy. He also gives an insight into just how much goes into guiding the attention of the audience in a subtle way for a particular kind of musical number:

> **Moran:** Lots of people have talked about the similarity between what we do and the roles of the director of photography and the editor in film. It's more than the working with the psychophysics of, 'attention is drawn to the brightest thing within the field of vision.' Are you conscious of that in your work? Of being the person who is directing the audience's vision?
>
> **Ben Ormerod:** Very, very much so, yes. It's a huge responsibility.
>
> Consciously, in something like in a musical, for instance, where everybody is radio miked. In *Zorro* there's a number where there are so many

> girls singing, they're all standing still, they're scattered around the stage, some of them at different levels. Each one has one line to sing. You have to make sure that the audience are looking at the person who's about to sing, otherwise, however great the sound design is, it's impossible to locate each voice precisely and your eye ends up wandering aimlessly around the stage. And, yet, you don't want stabbing spotlights.
>
> In the number, it's got to feel like people are standing in a single lighting state singing, and yet the audience have to be looking at the right person. Then this is actually where I might too have a word with the director. 'Could they just move, just a little shift, physically, in their body? Just to help?'
>
> Yes, so we've got [lighting] cues that build [the intensity on] the girl who is about to sing. But, you know, the cue has to happen before the person sings, and the placing of that cue within the musical score has got to be such that musically it makes so much sense that you don't notice it happening.
>
> So everybody is already lit, and then another light comes on to pick each one out a little bit better, cued within the musical structure so you're not aware of it happening, and because you're not looking at them anyway. And then just in time for each new line, in the corner of your eye, a little bit of movement of light that just takes your attention to where it needs to be.
>
> What's difficult about all of that, I think, and this is a real skill that a lighting designer needs to acquire, is then being able to watch that as if for the first time, and to be able to say, 'Does that work?'

Achieving this naive viewpoint is an important skill for the dramaturge too, and indicates something of how the roles overlap.

We went on to talk more generally about focusing the attention of the audience by controlling distraction:

> **Ben Ormerod:** I don't think about focusing attention very much; sometimes I do, obviously sometimes I do. Most of the time, it's about not being distracted by other things.
>
> A stage is a kind of receptacle of meaningful objects. Some of those objects are not always relevant, so you don't light them. Most of the time, the cues that are doing that kind of work are more about removing distraction than they are about revealing significance.

And then there are practical considerations, again driven by removing distraction:

> **Ben Ormerod:** I'll be watching a scene; I'll think, 'What's that chair? Why is that chair so brightly lit? That's ridiculous.' Often, you're finding that you'll have something like an actor in a dark dress, sitting on a light chair, you could spend your whole life brightening that area when they sit, calming it down when they leave and so on and so forth.

This assumes that the lighting designer is sufficiently inside the process to know where the focus should be at any given moment:

> **Ben Ormerod:** Yes, you know where you are supposed to be looking and then you just make sure that your eye doesn't wander to a brighter bit that you're not supposed to be looking at. It's done as a negative outline...
>
> **Interviewer:** Like drawing with a rubber?
>
> **Ben Ormerod:** Exactly, that's right. That's one of the things I learnt from Gerry Jenkinson. He would just spend ages turning lights off. Michelangelo said, 'When you're carving, the sculpture is inside and you chip away the marble until you can see it, and then you stop.' I think that's why I don't often do plotting sessions, because you're guessing at whether you've finished or not, whereas, when you've got actors on stage, the band starts playing, somebody starts singing, you turn a light on and it looks fantastic, so you stop.

Returning to Mark Henderson, watching the final run-through in the rehearsal room is most often his way into finding out what should be the focus of his lighting design at any particular point in the piece. Implicit in this practice is his having a clear understanding of the story gained from his own reading and research, and from talking to other members of the creative team:

> **Moran:** [Can we talk about your take on] the LD as dramaturge, as guardian of the story we're trying to tell. Is that something you see as useful are you conscious of [taking on that role]?
>
> **Mark Henderson:** Umm... I suppose I'm not over-conscious of it. I guess you are in a way because you are helping to tell the story. I don't think too deeply about it.

My own experience of Henderson's work is that light works really hard and well to inform the telling of the story in his lighting designs, and numerous critics and award judges seem to agree. We continued:

> **Moran:** Do you see it as part of the LD's gig to make sure the audience see those details that are important to telling the story, and doesn't get distracted by the wrong stuff?
>
> **Mark Henderson:** Yes. I guess it becomes about focus doesn't it ... it's a multi-layered thing isn't it because as well as focusing on the people you're focusing on atmosphere or the music. Whether you're conscious of it [or not] there's a double telling of the story.

For Henderson, then, helping to tell the story is important, and motivates a lot of his decisions in his search for the *right light*. But theatre is not solely concerned with simple linear narrative – as Henderson says: 'there's a double telling of the story'. Often much more than a double telling.

Lucy Carter talked about how the multi-channel expectations of our digitally enhanced lives feeds into the way we make and receive theatre, in particular some of the high-energy dance work she is perhaps best known for:

> **Lucy Carter:** The way an audience attends to something, because we are [all] so used to things shifting so fast and telling more than one story, they're looking for more than just the story. They're looking for the reference, looking for the psychological implication. I think there's so much in everyday life and in film ... that they come to theatre needing that almost over-stimulation.

This notion of multi-channel has similarities to Jon Clark's way of describing his contribution to the process of making a production as 'one voice among many'. Without a clear focus, however, the many voices can sound like chaos. For narrative drama that clear focus is very often the telling of the story. Here James Farncombe starts with the simple direction of audience attention through use of intensity, and then expands upon it:

> **James Farncombe:** I suppose when you said that [about prioritising story-telling] I automatically imagine the cue to close down to stage left and another cue open up on stage right and another cue to take focus up centre, and actually maybe it's to do with just making sure that you can see what you need to see and my job's about understanding what's necessary and what's not. And usually that's pretty obvious. And sometimes it isn't, and sometimes you have to interrogate the director a little bit to figure it out.

Interesting here that Farncombe says it is usually pretty obvious 'what's necessary'. These interviews have led me to believe that this kind of understanding comes from immersion in theatre making and forms a necessary part of the lighting designer's instinct. We continued:

> **James Farncombe:** Even in shows where naturalism is the watch word, there's room for that [focusing on what is necessary], and I'm not afraid to ask people to move to the light either. If you need to see stuff and you want to have that effect of just the room lit from the window, then put the people in the light from the window and play with it. You can move in and out of it, you can do all that sort of stuff and it is great fun when you get to the point where the actors are conscious of the light and are working with it. It's magic. So often they've got too much else on their plates to possibly think about the light. But the ones that are really good take all that on board.
>
> So I think there are ways of doing that [story-telling] that are subtle and subconscious [for the audience]. You can pull focus without specials and all that sort of stuff and then use accents occasionally, just to augment a particular face or a particular part of the stage, as long as it's part and parcel of the larger key and doesn't look out of place.

This next section seems to me to be key to an understanding Francombe's approach:

> **James Farncombe:** For my money, it's lovely when, whatever it is, [the light that is helping to tell the story] might have existed in that space anyway and it just happens to support the idea. It just happens to be a moonlight that comes through that particular window at that point. That's lovely 'cos then you really feel like you're conjuring a story out of the space that exists rather than imposing it on top.

The metaphor appears to come naturally out of the space, and the text. The lighting designer makes an opportunity to play with it – to engage with it, maybe even to expand it a little. As we will shortly see, Paule Constable is very much of the same opinion.

But that's not the only possible approach:

> **James Farncombe:** [A]nd then of course there's a show where it's completely appropriate to have specials come up and the language is ... [all about very obvious highlighting of one performer or one area of stage at a time]. It's about establishing the language isn't it? And that can only come

> out of conversations with designer and director at the outset. It would be crazy if you went and imposed your own language on every show you did. I'm not sure anybody who makes a living from lighting does that.

So the lighting designer uses light to help tell the story. This may sound rather more like traditional than integrative practice, though Farncombe and others are unafraid to arrange things on stage, including the performers, to allow lighting to contribute to: 'conjuring a story out of space' – something that a lighting designer working in a traditional way would rarely if ever have contemplated.

Working closely with a director

UK lighting designers, particularly in drama, are still reluctant to talk too much about the power of their light. If the impression some of these interviews give is that little has changed in the practice of lighting design on stage, it is perhaps because we are not being told the whole story. Mark Jonathan has a reputation not only for producing stunning lighting designs for opera and dance, but also for 'telling it like it is':

> **Moran:** Talking about the LD as dramaturge, not in the librarian sense – in the sense of guardian of the story, illuminator of the story practically and metaphysically. Does that work for you?
>
> **Mark Jonathan:** Yes. I'm older and wiser now and sometimes I know to just shut up and sometimes I know to, maybe, whisper something to the director. I'll show you the pictures of a wonderful cooperation between the director, designer and lighting designer. It was *La Traviata*,[1] we were doing a shadow play story in the prelude to Act 1 and the prelude to the final scene.
>
> I went out of the room for five minutes. When I came back they were talking about video projectors. I went, 'No! It's not a video projector'. Of course, it was going to be Bertha – the light would became a character with a name, called Bertha because it was quite big for the small scale we were working on – a Scottish Opera tour to the Highlands and Islands. It was a Cantata PC [an older tungsten lighting fixture] without the lens in it that would give really crisp shadows.
>
> I knew we were going to do something, we were going to tell some sort of story with it and I wasn't quite sure what. I'd shown the director, 'This is how big the shadow is when you go here and look, when you go away, it's

> the opposite. It gets smaller. We can have one person big and one person small,' giving her a few ideas to think about.
>
> Annilise starts directing the opera, and we had Bertha in the rehearsal room. What I did was I would lean forward to the director and whisper things. Then, she would very graciously say, 'Marko thinks that you should all get your banknotes out one after the other in time to the music' (laughter).
>
> That was up to her to do that. She could also own the idea if she wants and just say, 'try this' or 'do this.' When she thought it was a good idea she would tell everybody it was my idea. I was certainly beyond the strict world of lighting. They did all get their banknotes out in time to the music. Then, I whispered, 'Kiss them, kiss the banknotes so we'd know it was about love for money and then shower the money over Violetta.' It was breath-taking; it was fantastic to do.
>
> This was a director with whom I was on my third opera. So there's complete trust and a complete engagement in what I could bring to the table.

As Jonathan pointed out earlier in the interview, since a lighting designer works on maybe 12 to 24 productions a year compared to at most 4 or 6 for a director, set designer or costume designer, a lighting designer with 10 years of lighting productions will usually have done far more shows than a director with a 20-year career. Many value this experience in a variety of ways. Sometimes the lighting designer as dramaturge is called upon to make the whole thing practical. As we heard from Neil Austin in chapter 1, he frequently finds he is asked to choreograph scene changes. Here Jonathan talks about producers seeing his experience as valuable in helping a less experienced set designer make ideas practical:

> **Mark Jonathan:** Dramaturge, yes. When it's appropriate, yes, I enjoy it. I think it also comes with experience that you can't get out of the box. I think the clever producers are cross-casting. I know that I've been cross-cast with a Linbury Prize winning designer. They've been quite clear that part of my brief is to make sure that this production is practical.

In other discussions it was clear that Jonathan's definition of 'practical' goes beyond 'can it be built safely, on time and in budget?' to include, 'does it help to tell the story in a useful way?' (The Linbury Prize for stage design is awarded every two years, generally to designers who have very recently graduated from training, and consequently have little or sometimes no practical experience.)

Katharine Williams has a slightly more literal take on her role as dramaturge, particularly in her working relationship with one particular director:

> **Katharine Williams:** Oh yes. Alex Clifton – in the two plays that he directed at Chester, both of them he said to me, 'I don't know how to end it. What do you think I should do?' And I said, 'we should do this, this, this and this.' And both times it worked. And that's from years of – long before I ever knew what a dramaturge is – years of going, 'what we have here is this, shall we play with it in that way, or shall we change it like this'?

Perhaps more directors, and theatre writers, should collaborate more closely with lighting designers – but I suspect that many already do, without either party making any great fuss about it.

Williams has some useful and interesting things to say about collaboration in general a little later in the same interview:

> **Moran:** When you have been away for some of the rehearsal period, how do you go about getting back into the world of the piece?
>
> **Katharine Williams:** Well there are surprises, but within the tool kit [that I have planned to use] there is always a way of making it work. And I think this is a thing [I learnt from doing] devised work. The designers I know who have come through a devising background seem to be really hot on this.
>
> When a show is working as well as it can be, that's because every single person on that show is making the decision that is most right for the show. Not that's most right for the lighting or the sound or whatever their bit is, but for the show. The performers are doing that and the director is doing that too. And sometimes the director is taking a decision that is great for lighting and sometimes not, but that's ok. So when [you come back and] it has all changed – most of the time it's in a way that is more right for the show and you get back and it's, 'all right! Let's do this.' ... So I don't get fazed by that.
>
> I also think that when you have done enough shows where you make the show and then put it in a different order on preview two, I think when you've done that, you just don't get fazed any more.

The idea that when it is working well it is because everyone is 'making the decision that is most right for the show' is a lot like Clark's voices being in harmony, or Chivers being 'in the zone.'

Williams's points tie in well with Ormerod's notion of working with a structure, which we heard at the end of chapter 3. If the structure is right for the show; that is, it has been developed from within the production process rather than imposed on the production from outside, then even if changes in staging mean that the lighting designer needs to rethink plans for every lantern in her rig, the underpinning structure will still work. This is perhaps a marker for integrative practice and an active aesthetic, concerned first with what the light does rather than what it looks like.

Poetry and metaphor

One idea that runs through several of these interviews is the analogy between lighting design and poetry – in chapter 4 we heard from Jon Clark that at times for him 'it [light on stage] is not trying to stand out to be descriptive but it's melodic ... it has a rhythm and a melody in the same way that poetry does...'; again, not imposed from outside but an active aesthetic, developed collaboratively within the production process. In chapter 3 Paule Constable talked about how poetic metaphor in the text can provide the stimulus for metaphoric use of light. Here is Mark Jonathan on the same subject:

The weather on stage

> **Mark Jonathan:** So there isn't a rule – each piece brings its own rules, or...
>
> **Moran:** Language? Would that be a useful term?
>
> **Mark Jonathan:** Yes, yes. Let's take emotion. ... But it is funny how – in naturalistic terms – the weather [as described by the playwright, librettist, composer or choreographer] responds to what's going on on stage (laughs). How much we [as lighting designers] can do [anything to support] that depends on the set design. I mean if the designer gives us a black box, then we can't do all that cloud work on the sky – but if they give us a sky then we've got that opportunity.
>
> I do think that ballet and opera may allow us to express emotion more boldly, perhaps, than drama. But again it is very dependent on the piece.

Jonathan here uses the idea of establishing a language for each production, an idea I first came across in talking to Rick Fisher some years ago. But he also reminds us that the dramatist (or librettist for that matter) is using metaphors like the weather that allow the lighting designer to find motivation for metaphoric changes from inside the piece – inviting comparison to what Farncombe talked about: 'conjuring a story from out of the space.'

Around the time of these interviews, one of the boldest design metaphors I'd seen on a drama stage was the massive array of lightbulbs that floated above the expansive Olivier stage for the National Theatre's production of *Frankenstein*[2] in 2011. Bruno Poet was the production's lighting designer. Here he talks about that object and the production:

> **Moran:** For me, of the stuff of yours that I've seen, the most spectacular use of light as metaphor was *Frankenstein*.
>
> **Bruno Poet:** Yes. Well. You couldn't really miss that one.
>
> **Moran:** We were in the front row of the circle for one of the previews, and the heat from it [the large array of tungsten-halogen lightbulbs] was amazing.
>
> **Bruno Poet:** Yes, the heat of it was something else. We didn't expect the heat. It's obvious when you think about it, but we really hadn't thought about the heat.
>
> **Moran:** Where did the idea come from?
>
> **Bruno Poet:** When I first met [director] Danny Boyle, having read the script – the script talks about the creature opening his eyes, and blinding light – so we had to find a way of doing blinding light. Mark Tildesley, the set designer, found this image of a man in his vest and pants in a room full of lightbulbs. That inspired the idea of the ceiling of lightbulbs. It kind of grew. In the first model I think he had around 400 kind of crystal balls above the stage. …
>
> So we worked together until we found a way of achieving it. Mark drew it up, and we had sort of guessed it was about 600 to 700 lightbulbs in total. When he drew it up at the density required to make it work visually, it ended up being 3000 lightbulbs, maybe three and a bit thousand – I can't remember exactly now. So how do we find a way of controlling 3000-plus lightbulbs individually? [The Olivier theatre has around 1000 dimmers installed, which is considered a lot.] We went through all kinds of different routes – including fancy electronic lightbulbs – and ultimately it came down to cheap dimmer packs. Then we divided the ceiling into 1000 squares, and whichever lightbulbs were in that square became one pixel, and then we pixel-mapped it and controlled it like video.

What was great about that whole thing – I mean it looked spectacular – but what was great about it was it was a fantastic piece of story-telling. It did several things for us: it was a great scenic element – it gave a shape to the Olivier, which is this vast space; in terms of focusing the audience into the action I think it was very useful; and then the way we could do the flashes in terms of the creature waking up, the brightness and the sense of power and electricity – and the feeling that it was of the right period, the electricity of those early days and experiments – that sort of organic quality I think was very important. And also it became something we could use to punctuate the sound effects and have bolts of lightning. It became lighting in the sky, it became a metaphor for beauty – it worked in lots of different ways. And even un-lit it was a very beautiful thing, so when we got to the North Pole, it became icicles:

Moran: As you say it worked very well with the sound – tying sound and light together.

Bruno Poet: Yes, and when you turned it on it buzzed – because of all those cheap dimmers at very low intensity – which Underworld [the collective that created the sound design] sampled and just put round the surround-sound [system in the theatre] and manipulated so that the sounds you heard were 'played' by the ceiling, and manipulated and sent round the theatre.

Moran: Fantastic! That sort of level of collaboration across all the departments ... that must be exciting?

Bruno Poet: Yes, its fantastic. And I think it's very normal. I think that's what we do.

'That' may be what contemporary integrative lighting designers do, but it is not at all what traditional practice says it did.

Trying to get Paul Pyant to talk about poetic lighting

Paul Pyant's work is often referred to in professional reviews (and by students of lighting design) as poetic. He, however, is reluctant to call his own work poetic, though it becomes clear here that achieving a kind of poetry in the production as a whole is part of his aim:

Moran: The word that gets used sometimes to describe what I would think of as good lighting, is 'poetic'. I think that that's in the sense of the way in which poetry foregrounds language as its medium. It's important which words you choose in prose but it's a different kind of relationship and communication. I wonder whether you think that poetry is a useful analogy for what you're trying to do?

Paul Pyant: I never thought of myself as [creating poetry]. ... There are moments of true poetry that happen, it's true, but that's a good way of putting it I suppose – a bit prosaic and flowery for me – but ... It's always what's right, and whether that's poetic or not? It's about responding... Depending on what you're doing isn't it really?

Certainly, I would say, in the Shakespeare stuff I'm involved in, Lear at the National, or anything I've done Shakespeare-wise. Actually, if it's a monumental moment, the expression of the space by using light coming from a long way away, so you get the sense of some kind of infinity, I think that can be poetic...

When Hamlet comes in and talks about the seers or whatever, it is about that mercurial moment, where the text moulds with the actor expressing those words, and his delivery of them, in whatever space you've given them to do that, and I personally don't think I should impose too much on that. I mean I think that changes with directors and things but if you can give them a space...

There were terrific poetic moments in *Versailles*,[3] typically tragic things. The last image of the show – the show went backwards and you saw the 1914 scene having been through the 1919 scene, so basically the show ended with you knowing what happens to everybody, and it's knowing in that moment the enormous tragedy. But that was a moment with just four ladies on stage, and a bit of Butterworth [early twentieth-century classical composer who was killed in the First World War], and a long, slow fade and ... I was on the floor every night. So it can be poetic, absolutely. That's a good way of expressing it I think.

It has to be of the whole thing, I think. And when it works, it works. You've been involved in work where you've gone, 'Wow' even if you're at a rehearsal at 10.30 in the morning and you see that happen on stage and you go, 'Well that's pretty damn good'. But that does depend on so many things. It has to be the right play, the right words, the right actor in the right costume, in the right set, in the right light and...

Moran: What helps you decide what the right light is?

> **Paul Pyant:** Oh dear. Well again it's a gut thing really. I don't really know. There's many the time I'm sat in my studio on my drawing board, fiddling around thinking, 'Now what can we do here?'

In an effort to get Pyant to talk more directly about his contribution to making poetic moments on stage, I asked him about his work at the the Donmar theatre. It is a small commercial studio theatre with a reputation for very high-quality work. The Donmar's recent artistic directors include Sam Mendes, Michael Grandage and, as I write, Josie O'Rourke. Paul Pyant had lit 14 productions there at the time of this interview (2014) so I tried to ask him what he found interesting about making work in that space:

> **Moran:** What makes the Donmar work so well for you, because I've seen some beautiful stuff of yours there?

> **Paul Pyant:** Just the immediacy of it all, I think. And you can still be poetic in that space I think as well. That back wall has been dressed in so many different ways, and it's been reproduced in so many different ways it's unbelievable.

Pyant retreats from talking beautifully about poetic moments on stage to addressing some of the practical issues of his own practice. This reflects his instinctive approach to lighting design – an instinct built on many years of working with top drama directors. Where others find it easier to talk about their own lighting, Pyant's focus is always on the whole effect; the poetry made when every element of the piece is 'right' – and this too is an important point. Lighting design on stage does not work in isolation.

The right light as poetry?

Rick Fisher is more forthcoming on the contribution light can make in its own right, though he too mentions the general reluctance to talk seriously about light's poetic role on stage in the UK, even amongst other theatre makers:

> **Moran:** One of the other theories I'm trying out is around poetry and prose. Language draws attention to itself in poetry – there are times when lighting draws attention to itself...

> **Rick Fisher:** It is an interesting analogy. I've never really thought of it that way, but it is not a bad analogy. And if I ever used it in a production meeting or a tech session I'd probably get kicked (laughs). But – yes, sometimes the lighting has to be there, like a warm bath – making you feel

good without you realising it's the water and the temperature of the water that's doing it for you. And sometimes the lighting has to be something like a wet fish in the face, and smack you into another thought. It's knowing when each one is right, and sometimes you're surprised.

There are classic moments, when directors will often say to you – or writers will – something like, 'right. This may be a wet fish moment. Last night I had a dream...' – that lazy kind of writing when we need a bit of exposition that we can't justify in any other way – or, the telephone call, to give you information, or the director saying: 'at this point I want to go inside their head'. How often have you heard those things? Or: 'this is a moment that's out of time'; and you think: 'OK. This is a moment when lighting is more poetic.' But I do find it is useful, even for someone with 30 years of experience, to somehow ask the director, or the designer, at some point – and in words that are fairly clear and understood by people who are not lighting people – 'does the realism of this moment work? Is it necessary?' ...

[B]ut I think there are possibly less artistic sounding words for poetry versus prose that might be useful ways of discussing some moment with your creative team. You know, 'is this naturalistic or expressionistic?' And sometimes you just have to show people, 'it could be like this or it could be like that'.

Of the lighting designers interviewed here Paule Constable is one of the most comfortable talking about her work as art. Towards the end of our interview we talked directly about *rightness* and the vulnerability you have as an artist, particularly making art in the public space of a technical rehearsal, echoing a point Mark Jonathan made graphically earlier:

Moran: So would you like to sum up what makes the light 'right' on stage for you?

Paule Constable: I think purposeful and beautiful. To be poetic – which means choosing the right 'word', and leaving out all the ones you don't need – that makes it right, doesn't it? And god its good when it's good isn't it?

But it's so elusive. But keep striving for it, that's the thing – Quite a young lighting designer I know well, I went to see a show he designed and didn't know what to say because the lights weren't great. And he said: 'well I was doing another show at the Hampstead and this was done in an afternoon'. And it looked like it. It is not good enough to do that.

> **Moran:** R.G. Collingwood was an Oxford professor writing about art between the Wars, and of course he is writing in the wake of Duchamp and, 'I am an artist. If I say it's art, it's art.' And he's saying that what lifts the work of a jobbing portrait painter from the level of pot-boiler to art is the painter's investment of emotion in the work...
>
> **Paule Constable:** The problem with working in that way is that it makes you very vulnerable as well. Which I know. So there are repercussions of that ... in a way you can see why people end up doing it 'at arm's length.' But for me – if you're going to do that [work at 'arm's length'] don't do it. ... No heart – no art!

Constable here is very comfortable with Collingwood's notion that art requires some emotional investment from the artist. Perhaps if a particular production is art, it is art made in collaboration – requiring high levels of trust between all the makers as well as from the performers, each one voluntarily exposing their vulnerability, and risking exposure.

'Living inside the metaphor'

These discussions reveal a practice that is concerned at least as much with what light on stage does as does as what it looks like. It seems to me that this is one important way in which the analogy between poetry and lighting design works in narrative theatre. In most narrative poetry the subject drives the choice of words – of course the more right the words the more right the poem, but it is generally that way round. Similarly, in simple terms, in lighting design for narrative theatre, the intention of the lighting designer generally comes first. What do you want the light to do? This is what most often drives the choices of which volumes of space have what quality of light – and darkness – for each moment of the performance. After that it is partly a technical job to create those volumes of light and darkness. This is not to say that this is any more trivial than choosing exactly the right word for every beat of every line of a poem, or that it is wholly technical, depending as it does on aesthetic choices made by the lighting designer and a sensitivity to the flow of the piece that will inform every cue state and transition.

Ben Ormerod has an interesting and revealing take on how his minutely researched understanding of the form and structure of a piece – and thus the lighting design that will be part of that piece – informs his creation of the content of that design as he makes each cue in the technical rehearsal. He calls it 'living inside the metaphor':

> **Moran:** Paule [Constable] made a really interesting point; that light too often is pedestrian, and it's because it's too easy to just put some light

there. The whole thing of making that conscious, informed choice, in your case based on the structure, begins to take the light into the realms of poetry.

Ben Ormerod: So you've got this thing, haven't you, this thing that we call 'Metaphor', which is a really interesting phenomenon. This is where I think light and poetry are associated somehow.

On the few occasions that I've directed I've been interested in finding a way of helping actors deal with verse as characters. Students in particular have tended to either attempt to make it sound naturalistic, as if they're talking about banal everyday things, or make it 'poetic' – which generally means putting on funny voices. So I have suggested a different way of regarding the use of poetry on stage; that instead of living in a literal world and describing it poetically, the character should live in a poetic world and describe it literally. In other words, when Macbeth says, 'What is the night?' and Lady Macbeth says, 'Almost at odds with morning, which is which', he's not saying, 'What time is it?' and she's not saying, 'Half past five in the morning'. He's saying, 'What is the night?' and she looks out of the window, sees the day and the night in conflict, and describes it. This is much easier than finding the motivation to say 'it's half past five' in a peculiar way. I call this 'living inside the metaphor'.

My approach to lighting poetically, then, is to say, 'This is a metaphorical world that we're creating on stage and my job is to create those physical objects that metamorphose'.

In *Tristan und Isolde*, for instance, I've got an interesting problem. Sunlight, which for so much of my career has been this Apollonian, healing presence, represents in Tristan a world of appearances, of the deceptive persona, of glamour. And so where in *Macbeth* the night is the world of darkness and concealment where evil flourishes, in *Tristan* the world of night is the world of truth, the world of lovers, where the personas are stripped away and truth and honesty reign. That's interesting.

So at the moment, one of the things that I'm thinking about with *Tristan* is how my relationship to sunlight, which has always been benign, speaking of truth and love, has to be revised. What should that look like? (There I am, saying, 'I don't concern myself with what things look like', but obviously, I do.)

But my point is, that's now, in preparation. By the time we get into the theatre, my job is just to make sunlight and moonlight, I'm immersed in the metaphorical world and I'm not actually thinking metaphorically any more. The lighting is living inside the metaphor. 'Okay, we need to introduce some sunlight, because here comes Kurwenal'; 'Okay, it's got to be night here, because Isolde's extinguished the torch.' In my preparation, the torch represents sunlight and danger, as well as fire, but in the tech, I'm immersed in the metaphorical world and I'm just doing everything literally; although unconsciously, of course, who knows what's going on...

Moran: In that way, you remain true to the text, as opposed to trying to put your own spin on it?

Ben Ormerod: Absolutely. And so there are no funny voices in the lighting. That's key.

Making practical and technical choices from inside the metaphor – perhaps this is what Chivers means by being in the zone. Having developed an instinctive understanding of the metaphors and structure of the piece, the lighting designer makes their informed, technical contribution to telling the story.

Summary

For our lighting designers, then, a focus on the big picture, combined with the 'fresh eyes' that come from standing slightly outside the intensive crucible of the rehearsal room for at least some of the process, underpin the importance of a good dramaturgical understanding as part of the lighting designer's necessary instinct, particularly for narrative theatre. The 'fresh eyes,' however, are not as naive as those of an ordinary audience member. They are informed by reading and researching the text, and generally by the discussions had at the beginning of the process (though recalling Pyant's comments about working with an older generation of directors who don't really talk to him, this may not always be the case). And the lighting designer can ask questions. But on top of everything else, as Ormerod says, the lighting designer has to find some way of being able to 'watch ... as if for the first time, and to be able to say, "Does that work?"'. And if it does not, be able to take it apart and fix it.

6 Dance and Abstraction

This chapter focuses on lighting for dance, and particularly on non-narrative dance. To quote lighting designer Michael Hulls from an interview with dance critic Judith Mackrell in February 2014, dance is 'where light can reach its full potential'. Later that year Hulls won an Olivier Award for Outstanding Contribution to Dance. Here is some more from Mackrell's article:

> **Peter Mumford:** 'I really think dance has led the way in exploring and expanding what light can do on stage ...'
>
> **Michael Hulls:** 'Light enjoys a freedom in dance, especially non-narrative dance, that it can't have in opera or theatre.' In fact, Hulls believes dance and light enjoy a special kinship, thanks to their shared ability to dramatise space and convey nuances of emotion. 'They're both so hard to talk about,' he says, 'so ephemeral. They don't exist except at the moment of performance. That's why I only work in dance. It's where light can reach its full potential.' (Mackrell, 2014)

Michael Hulls: Icon of contemporary lighting for dance

Hulls is an icon of the UK contemporary dance world, but less well known beyond it – in part because of his refusal to work outside his favoured genre. Here is what *Telegraph* dance critic and founder of The Arts Desk.com Ismene Brown wrote in her article following Hulls' 2014 Olivier Award:

> Lighting designers are either wizards or useful pedants. They scrupulously light the action or they make light speak its own language, activating space, time, illusion, imagination – inventing effects that your blinking eyes can only consider as magic. No one performs this wizardry more outstandingly than Hulls, who has become a co-creator in a new form of dance-theatre, where light and movement are an inseparable duet.

> He and Maliphant invented this collaborative idea tentatively 20 years ago, and their unbroken partnership ever since has led from obscure early solos to the international acclaim brought by their creations with Guillem and the Ballet Boyz, the astounding trio at the Royal Opera House *Broken Fall*, and Akram Khan's astonishingly beautiful *DESH*. Only just this last week Hulls, while admitting on Twitter he had not had enough time to do something spectacularly original, contributed the aching between-worlds atmosphere of the new Maliphant work for English National Ballet in their World War 1 new-choreography programme, *Second Breath*:
>
> Hulls' list of achievements is a catalogue of many of my most unforgettable experiences of watching dance for the past 20 years. It was he who bathed Guillem and Maliphant in auras and glows in their duet, *Push*, and in their picturesque orientalist work with Robert Lepage, *Eonnagata*. It was Hulls who brought out with his lights, in Maliphant's duets for the Ballet Boyz, a powerful undertow of both male rivalry and unbreakable friendship. A recent production cited by the Oliviers last night was his work with the Ballet Boyz's latest double bill by Liam Scarlett and Maliphant (Brown, 2014).

Clearly, in an abstract dance piece light is not going to be servant of the story. In our interview Hulls makes it all sound quite simple, but then to quote Bryon again, 'the best choices are not complicated; however there is a certain level of complexity involved in the planning' (Bryon, 2014):

> **Michael Hulls:** Generally I want to work in a black void because then things can disappear more.
>
> Jennifer (Tipton) says, 'Put the light where you want it, take it away from where you don't want it,' which I understand to be about the quality of darkness, which is almost – almost – as important as the quality of the light because they are opposite sides of the *same* coin.

Even on large stages, outside the world of classical ballet, dance tends not to rely on set in the same way that drama usually does:

> **Michael Hulls:** Another generalised but huge difference between theatre and dance is that in theatre often the visual work is determined by a set designer, and in dance there is often no set, or very little, and the visual world is determined by the lighting designer or in some cases by a costume designer.

Dance lighting practice informs contemporary drama lighting design

Several lighting designers have commented that a big factor driving the change in style from traditional to a more active practice is the influence of contemporary dance lighting with its focus on, for example, activating the body in cross light. Peter Mumford's fascination with light (when he was training as a set designer) was triggered by student projects working with the company now known as Rambert Dance. He went on to work professionally with them:

> **Peter Mumford:** I discovered [a lot that I've taken into my other work] with the dance stuff I was doing in the 80s, particularly with Rambert, when I worked with a lot of painters – I worked with Howard Hodgkin, worked with Richard Smith, I did work with a whole load of them. They were all people who brought a kind of a naive thing ... to the theatre – because most of them didn't know how theatre worked, but they knew how to make images. And there was a tendency in some instances for some lighting designers to be rather afraid of them and think, 'oh it's a piece of fine art I have to put white light on it', but I didn't do that. I messed with their colour, and changed their paintings, and they loved it! It was great – they totally entered into that.
>
> So you could take that, and you could take it on further and say: 'well I can make that red disappear'. That was really good in terms of learning about input, and what the lighting does to tell the story of the piece, even quite an abstract story. It has a huge role in that sense. I would take that attitude into everything, from the most naturalistic of plays to the most abstract of dance pieces. It [light] does fulfil this almost cinematographic role of story-telling as well.
>
> **Moran:** You talk a lot about surface and painting. But of course we work in a three-dimensional space, particularly with dancers. Is sculpture a similarly useful idea?
>
> **Peter Mumford:** Absolutely. Yes. Lighting is very sculptural. ... If I'm working in an empty space – like doing a dance piece where there is nothing, just light and the performers – that would draw me more towards using the light sculpturally, because then it becomes the set if you like. One of the first things I did like that would be a piece called *Strong Language*,[1] which I did for Richard Alston on an empty stage and the light is very sculptural in that. Because it is in a black space, that's

> how the light works, rather than in a painterly way. … To work in a painterly way you kind of need to have a surface.

The idea of sculpting space with light comes up in other conversations about lighting for dance, as we will see.

Starting points – 'some strange place in my head'

Without a formal text, and sometimes without music either, the choreographer, the lighting designer and other designers, have to find other stimuli for creative starting points. The way Michael Hulls talks about this has some similarities to Mark Jonathan's ideas in chapter 4:

> **Moran:** Lots of people who do mostly text-based drama have talked about how you go to the text first. So where do you go?
>
> **Michael Hulls:** Some strange little place in my head, which is that creative part that comes up with the response to something. It's not text, it is kind of the result probably of various disparate creative impulses, or that they moved like that, or it sounded like that, or I liked that light on that drape in front of the window. Somewhere in your head these things may or may not make a connection.
>
> A dancer years and years ago wrote about it, and they called it, 'the process of mentation' [defined as the process of thinking especially carefully]. They were writing about where that creative place is.
>
> It is about making connections between various things that may not apparently seem connected. So that is cerebral, but for me it is intuitive. I work intuitively, and what that means is that if I am seeing someone dance I kind of have to personally get off on what I am seeing … in order to be inspired and excited, and interested – to be able to have a creative response. Which means that there are some people I am interested in working with and some people who I couldn't possibly work with, because I just see them do something and it doesn't touch me, it doesn't stimulate me, there is no spark set off in me.

As the interview with Hulls progressed, he outlined two contrasting approaches to initiating work. The process he has with Russell Maliphant is well known in the dance world, and we will look in some detail at that shortly. First, here is Hulls talking about a quite different way of responding to choreography, from the making of *Stop Quartet*[2] with Jonathan Burrows, mentioned in Brown's article above. Made originally in 1996, it is a piece that is frequently revived:

> **Michael Hulls:** So I had had this conversation with Jonathan [Burrows] ... and then he asked me to come and light a new piece that he was making, which was going to be the *Stop Quartet*. I watched rehearsals and I got video from rehearsal, and was watching the video at home and I still didn't know where the design was going to go. The dance is about pattern and rhythm at quite a complex level, over my head kind of complex. I am watching them dancing in a studio and I know that there are complex patterns and rhythms going on, and I realised that there was a pipe on the wall behind them, and a bit of a mark making a horizontal. These two things, a horizontal line and a vertical line, if I watched rehearsal without those things it was harder for me to see what was happening. Put a complex pattern against a very, very simple pattern – then for me it became more understandable.
>
> It took me a while to realise that, and then I did, and then that was okay. So the design has to be a pattern of its own that is a very, very simple mixture of horizontal and vertical that doesn't really change. [The piece] started off as a solo then someone else joined it and it became a duet, then someone else and it became a trio, then someone else and it became a quartet, so all the time its complexity is increasing and increasing – so to put it against a kind of Mondrian-esque series of horizontals and verticals gave me fixed reference points in space against which their movements and their patterning more apparent to me.
>
> So that was the thing. It is about recognising something of that kind. 'Why is it making more sense when I watch this bit of video from this rehearsal, rather than that? Ah, it's because of those lines on the wall behind. They give me a frame of reference, that doesn't change.' So that was about recognising something. Often the answer is there in front of you, but you have to be able to recognise that.
>
> A critic wrote about that piece [*Stop Quartet*] that the lighting design exactly complemented the dance. I suppose that, when I was thinking, 'What is the right light?' that was an example I thought of. That someone saw it in that way, that it exactly complemented what was physically going on ... all the elements were about patterns. So that was what was right about it. It was that it exactly complemented and in some way gave you a framework through which your appreciation of another complex pattern was facilitated.

Moran: How much, if at all, did the choreography change once that [light] became the design motif?

Michael Hulls: Not at all. One of the other things that we had talked about was about pattern and structure, and that the lighting would have its own pattern and structure which would be entirely independent of the dance (as opposed to my collaborations with Russell which are all about the inter-dependence of the light and the dance and its pattern and structure) ... What was important was that everything had its own independent rhythm and pattern, so there were no cues to do with, 'Because they do that – this happens,' or, 'Because of that sound – they do that and the consequences are that.' [The music, the movement and the lighting] are like three entirely separate, independent layers that kind of slide around a little bit in time, on top of each other.

Moran: But the changes that happen in each of those layers are not arbitrary?

Michael Hulls: No. They have their own independent logic to them, and rhythm. That is why I am working in dance, because there is the opportunity for it [light] to assert itself and have its [independence] – albeit it wants to be appropriate and add rather than subtract. But it has the possibility in this non-textural, less lyrical, abstract world, to be whatever it wants to be.

Light as starting point

Michael Hulls: The other way [that I like to work] is the ideal way that I would work with Russell. Ideally on day one of creation we are in a space with lights and I can say, 'I am interested in this light.' Russell can say, 'I am interested in this movement,' and we put them together and something either happens or it doesn't happen. If something happens, then you know that there is something to pursue there.

It is trial and error, to a certain extent. It just might be completely accidental or you just try things out.

So *Two*[3] was discovered in that way, from trying something out and we thought, 'That flaring of the arm and the blurring of it is what is interesting about this. These other things, the fact that it is happening across the stage with two railroad lines of light. That is kind of irrelevant. The interesting thing is what happens with the arm going through that light.'

> So we think, 'The moving side to side, that is not important. It's the arm through the light that is important. So instead of having two strips, one front and one back, let's maximise that and have four strips and make a square. They don't need to go anywhere.'
>
> So we made a two-metre square with a half metre band of light around it, and so what we are also doing is creating a square of darkness in the middle. Then we think, 'Okay, this square of darkness and this strip of light around the outside edge, this is where we want to arrive. So how do we get there?' We will start with a square of light and, over the course of several minutes, it is going to change into a square of darkness with the light just around the outside.
>
> So we discovered something through having time in a space with light and dancers, we both saw something and thought, 'Wow, that is exciting, that is something we want to pursue.' Then the [full] lighting design happened next. Then we put Dana (Fouras) in the light and Russell was happy as Larry, choreographing away *in the light* [author's emphasis]. That is his ideal situation, to be able to choreograph in the light.

Hulls went on to talk about how this extraordinary and productive collaboration has developed:

> **Michael Hulls:** When we first started talking about what we wanted to do to make the collaboration between light and dance, one of the ways of articulating it was that we wanted, instead of the traditional thing of light coming last in the creative process when everything else is done and dusted, then you have some lighting time and then the show opens, was to put lighting at the beginning of the creative process to kind of supplant the natural order where the primary collaboration is with the composer. (That is where it starts in a lot of dance, or because you have taken an existing piece of music.)
>
> It is different every time we work together, but the way that *Two* was made was kind of ideal. So that is through a practical trial and error based, 'let's get some time in a space with some lights and a dancer, and try some things out.'
>
> Sometimes something brilliant happens, sometimes you kind of go, 'okay, but not that interesting. Let's try something else.' Sometimes what can happen is you go, 'That is fantastic, but it is 30 seconds and there is nothing else we can really do with it, so we can't develop that into a "piece"; it's only a self-contained moment. It's brilliant but we will just move on from that.' That does happen.

> The other thing is to avoid coming up with something which looks like it is a demonstration of something. That we are demonstrating, 'I am moving my arm like this and the light makes it look like this.' [We don't want it to look like] we are demonstrating perceptual phenomena. That is not very exciting either. So there are things that are like a demonstration of a trick or something that is not going to give us the potential to make something that people can engage with.

Working with projectors – Afterlight

> *Afterlight (Part One)*[4] was an example again of some R&D time with some projectors and an animator. We were trying stuff out, which was graphic and linear and, 'It's not happening. There is some interesting stuff but nothing really going on.'
>
> Then, at home I'm looking at just a pendant fitting, a blue 60-watt bayonet lamp in it. A little bit of gap where the blue paint finishes just before the bottom of the cap, so strange white light coming out but blue light going up onto the ceiling. Then a swirly bit of really soft diffused textured white light is creeping out through this millimetre gap, a dark hole in the middle because of the cap for the lamp making its own shadow. I had been looking at this for years thinking, 'How can I make a light like that?' Lots of schemes about gobos and rotation and things and, 'No, it's not, it's not.' We are trying something out with video projectors and it's not working, and I think, 'That idea, plus video projector overhead.' I asked [our animator] Yan: 'Imagine how a cyclone looks on a weather satellite; the dark eye of the storm and then this swirl of rotating cloud that is organic and textured around it.' I got Yan to knock up this texture like a cyclone, from overhead. I asked [one of our dancers] Daniel Prior to get in the middle of it and start moving, and this thing is rotating slowly.
>
> So we tried Daniel rotating at the same pace as it, and ... with or without the light in the middle, and because [the projector] is directly overhead his shadow is on his own feet. So you cannot see his feet, so you cannot see that he is moving his feet to rotate himself, so he just starts to look like he is on a lazy Susan or something. So we're, 'Wow, that is fantastic, there is huge potential there.'

It's important to distinguish here between the thoughtful process of Hulls and Maliphant or Hulls and Burrows, and someone making arbitrary responses with light to a choreographer's work.

Lucy Carter's starting points for dance

Like Peter Mumford, Lucy Carter came to lighting design through dance. Though she now works outside dance, she is perhaps best known for her long-term creative collaboration with Wayne McGregor, which has resulted in some spectacular and award-winning work. She too talks about resisting the arbitrary, and puts in independent research work to ensure she is making an informed response:

> **Lucy Carter:** I think my starting point for every show, whatever medium it is, is different. I have to begin at the core of the piece. Of course with drama you get a script but sometimes on a devised theatre project you don't. I want to go away and research, just as a director would do their research. They tell me what that is and I'll go and look at that but I kind of go off-piste and try and find some inspiration, not just to be different but to find something that speaks to me rather than speaks to them. I always try and begin with a concept and go away and research that before I get too involved in the script, or the score of the music or what the story's actually saying sometimes.
>
> **Moran:** Do you document that research?
>
> **Lucy Carter:** I write notes. What I'm looking for while I'm researching is some inspiration that gives me a visual idea. Then I know my ideas have come from a core route and concept. They might be forming much later on in the process, I might be looking around for something that surfaces in a play and I would remember some research that I would have done. It's not massively in-depth but it's a few days' worth of time Googling and reading and looking, so I've got some roots to what I'm trying to do.
>
> **Moran:** So you want to get involved in the process as early as possible?
>
> **Lucy Carter:** Yes. Definitely I want to be in as early as possible. [Preferably] from the beginning. It's partly because I've come from a very dance-orientated background. Most of my beginning five years at least was dance, almost all abstract dance as well. This research that I do is to try and find ideas from nowhere because in abstract dance you're not portraying a place or a time, you're portraying a feeling, an emotion, and trying to direct the eye a bit like a film editor or something. So I'm

pooling my research into something with a choreographer, director or a set designer, meeting with their research and with their ideas.

With dance and ballet, certainly with Wayne McGregor, I've got a long-term [creative] relationship there, I'm there at the beginning. Then everybody goes off on tangents and develops bits together, just like Wayne and I might develop something without the set designer there.

Throughout the interview, Carter's fascination with light is very evident, and this comes primarily from her experience in dance:

Lucy Carter: It's hard to talk about light, because I often say to people who don't really know theatre that I'm lighting the air, you don't see light until it hits something. I'm also very interested in what lighting does in the air before it hits something. I guess in terms of dance, a dancer gets in the way and that's the interesting bit, the dancer is the air. It's disappointing that then it [the light] has to continue on somewhere and hit the floor or the wings…

What I realise now is that I'm getting better at talking about it. I used to never be able to talk about it. I couldn't explain what it was I was trying to do, so I always had to show in a way. You learn tools for explaining to directors what you mean; sometimes it isn't quite what you mean but you know that that's what they'll understand. It's super hard to explain something so ethereal and knowing it won't become anything until it hits the thing that it's meant to hit, and also it hits a performer in one way and then the set behind it another. So you've got to mould those two.

I guess I'm making imagery. I'm not making a series of photographs. Each cue for me is a new image in a book and then they're linked together.

Moran: [Those links – the] transitions – are you particular about the way in which one picture moves to another?

Lucy Carter: Yes I'm interested in it not just being about the cross-fade. In life you rarely cross-fade; you walk into a room and it changes. The environment's changed instantly and you turn on a light and it flickers on, or your eye takes time to adjust so it sees something in one colour and once it's adjusted it's leaving the room differently. I think that all those subliminal points between cues are really interesting, subliminal points between different environments in life are really interesting, so I quite often try to play around with those in the transitions.

> I suppose coming from dance, the transition is a whole section in itself... Also, as we all know, if you get into a cue in the wrong way the cue can just feel wrong, even if it's right.

I asked Carter about her process with Wayne McGregor's company, and how that compares to other dance and opera companies she works with:

> **Lucy Carter:** When Wayne [McGregor] and I make a piece with his company where we have full control, we have two weeks in a venue trying it out. We'll do a section and we'll say 'yeah that looks great, but shall we just try another option in case it's better?' You really get to mould it and I get to say, 'let's stop for half an hour while I re-work that'. So for dance [it is easier for me to get the time I need]. Maybe I'm just bossier [when I'm working with Wayne] but there's less of a technical team and a stage management team to call the shots [with Wayne].
>
> Once you get into theatre you have all this support team behind you, but the techs do stop for you to get the light right. Hopefully you have pre-programming time where you programme some scenes, so that's great. But opera and, more often than not, ballet [as opposed to dance] – you rarely get any pre-programming time and you do it all over the top of what's going on on stage.

Lucy Carter on light art

Carter has been an established practitioner now for some years. Unlike some other lighting designers of her generation and in common with Hulls and Mumford, she is increasingly comfortable with seeing herself as an artist. Indeed, when I spoke to her she had just made her first piece of light art sculpture:

> **Lucy Carter:** More and more I realise that I am an artist and less of a technician. ... I'm having more and more ideas that aren't from somebody else's first idea. That are just ideas about light and environment. I am exploring and I'm going to continue to explore.
>
> So now I'm starting to realise that possibly, yes, I'm an artist, a light artist, not just a lighting designer. I've just done an installation work in a castle in Switzerland for an art and music festival – which is certainly light and light object. It has a kinetic element and has a programmed duration of 17 minutes.

I mentioned R.G. Collingwood's idea of art requiring an investment of emotion, but it was the notion of art requiring heart, hand and head that sparked this next response:

> **Lucy Carter:** So [my installation] is in a castle in Switzerland and [an engineer] built it with this young lady who is a stonemason. She helped him construct this thing, and then I came in and decided how it is all rigged, and actually physically programmed it, but she then stopped me and said 'do you always do work that you don't build?', because she was a stonemason so she does it with her hands and it got me thinking very much about 'what is art?' Am I less of an artist because I'm not physically building the finished product?
>
> **Moran:** But that would rule out Damien Hirst and Rodin, and a lot of others!
>
> **Lucy Carter:** Totally! And a lot of conceptual artists! Then she asked me 'what do you think art is?' and I said I thought art was anything that stops you in your tracks and makes you think. Something visual, or aural, that stops you in your tracks and changes your day, even slightly, or changes your perspective. Making you feel something, because actually the stuff that I do [as a lighting designer or as an installation artist] I feel that a viewer, an experiencer, an audience member, doesn't have to 'understand' what I'm trying to say, it just has to move them in some way. Change their thought in some way…
>
> There's those [designs] where you're being totally subliminal and you don't want anyone to notice the lighting has changed or it's the lighting that's making them feel anxious, along with the sound. You're working on many different levels when you're designing like that. …
>
> [When lighting is not good] you know as an audience member who doesn't understand lights at all that something feels wrong, or you can't see or can't tell who is speaking. You know there's something making you feel wrong. Or maybe the show's not even working because you've got the wrong type of moonlight, the wrong coloured moon!

This strong image – the world on stage feeling wrong because it has the wrong type of moonlight – speaks so well to the lighting designer as dramaturge and maker of aesthetic choices, and to the responsibility of the role too.

Unlike Hulls, Carter splits her time fairly equally between dance, drama and opera:

> **Moran:** Moving between dance and drama as a lighting designer. Does that transition feel fairly natural?
>
> **Lucy Carter:** Yes, it does feel natural. ... If the right people ask me to do stuff then I'm not being asked to do super-realistic, naturalistic designs.
>
> I get very excited when I get a script because it feels very supportive in comparison to abstract dance where you're all learning together and it comes together very last minute and you're having to have quite a lot of ideas. Whereas when you have a bible, a script, it feels very supportive. You can go back to it all the time. You know ultimately as a creative team you're trying to portray that text, tell that story or portray that message. I think you're all moving in the same direction but dance can go off on tangents very often and it can change a lot at the last minute, whereas with a script you know you've got to tell that.

Dance lighting with larger companies

Some of Peter Mumford's most celebrated dance work of recent times has been with the Birmingham Royal Ballet company, with whom he made the suite *E=mc*^{2} (Einstein's formula for the equivalence of energy and mass), the design for which was a highlight of World Stage Design 2013 in Cardiff (Burnett, 2013):

> **Peter Mumford:** Yes, that was quite interesting because David [Bintley] said, 'I don't want a set I just want space. It has to be light orientated – do what you want' kind of thing. And in fact we did *Faster* like that as well (for the Cultural Olympiad in 2012).
>
> And you end up (being credited as) set designer on a thing like that, but I think what I'm doing is creating a kind of canvas for light. I'm thinking of both at the same time. I mean *E=mc*^{2} is a sort of set design and a light design as well. It does involve moving bits (stage flats and drapes, etc.) to make it work.

However, larger companies such as Birmingham Royal Ballet don't offer the kind of time to experiment and develop ideas with light in the way that Hulls and Carter relish. The large ballet companies, particularly the ones that dance with a live orchestra, are not able to stop their stage rehearsal for the lighting designer to fix something; there are simply too many people to pay for them to be doing nothing. (Much the same holds for large-scale opera, though they do tend to do more stage rehearsals with piano when stopping for technical reasons is slightly more acceptable.) This means that the lighting session before the dress rehearsal – without the benefit of the company on stage – is the only opportunity to try something out, and the lighting designer won't know if it has worked until the company arrive – and then there may not be time to change it. Added to this, it is not unusual for there to be only one lit rehearsal of a new work before that work is put in front of the public.

This may sound like a daunting and potentially unrewarding way to work; however, both Peter Mumford and Mark Jonathan have contributed stunning lighting to visually adventurous work of this kind, for example for Birmingham Royal Ballet. Here Mumford briefly talks about one approach to this way of working:

> **Peter Mumford:** I think on a dance piece, particularly if you're working with a company such as Birmingham Royal Ballet who don't really do lighting sessions with dancers, then you do have to go in with a clearer visual concept. So for example on *E=mc²* I created computer drawings, and I did the same thing with *Faster* for David (Bintley), to give him an idea of the kind of look. It's a tricky and slightly dangerous thing to do with lighting because if you start trying to do drawings of light then people get the wrong idea – or they get their own idea – and then it doesn't end up looking like that. So I try and avoid it. But I've found that because I was doing quite sculptural things with it I could make computer drawings which indicate what it was going to look like. And they were quite nice little drawings as well so that was fine. I think that did help David, and in that instance I did have a very clear idea of the look I was going for. … I do have a very clear palette to work with, and ideas in my head.

It is perhaps worth noting the contrast between what Mumford says here about working with a large dance company, and what he says about his work in drama: 'If the DSM comes up to say: "how many cues in the show?" I say "I don't know". And I don't…' But in this kind of dance, the production process more or less demands that the lighting designer is reproducing on stage ideas that have already been more or less completely worked out in their mind's eye.

That said, even in these tightly constrained production conditions there is still the potential to discover by accident something wonderful the light does on a set, providing you have done enough preparation.

Here Mark Johnathan talks about creating a 'coup de théâtre', again for Birmingham Royal Ballet:

> **Moran:** Taking a very specific thing, one of the most famous images of your work is the shot from the opening of *The Protecting Veil*.[5] Beautifully shot [by Bill Cooper] – and actually even better when you see it in performance.
>
> **Mark Jonathan:** Thank you.
>
> **Moran:** How much of that – because you've told me something about the process of creating that, and how little time you had in the space to create it – were you surprised by how it looked? Were you confident [it would look as stunning as it did]?
>
> **Mark Jonathan:** No. I wasn't confident at all. David Bintley [the choreographer and the company's artistic director] said, 'you've got two coups de théâtre: one is down to the set design and one is the lighting'. And I would have these nightmares, where I would wake up in the middle of the night with David Bintley saying, 'I don't like your coup de théâtre!' (laughs). And of course the first one is that photograph, which captures the lights not moving. And there was a reviewer who clearly reviewed the picture saying: 'Mark Jonathan's ever-present lighting rig'. Well of course it wasn't because it all flies out of sight. But the other coup de théâtre was the moment of resurrection. And in my mind he was going to be revived by the fingers of God – and I was going to do this with light. But of course what actually happened was that just before I got to that cue we ran out of time [in the lighting rehearsal] and they said: 'OK, time's up now we're going to do the run'. And I knew that I'd only just got to this second coup de théâtre. No one had seen it except me and the electricians who had focused it. And when we got the point [in the dress rehearsal] when we'd run out of cues, and I went, 'go on, put on that channel'. And it went 'kerpoing!' and everybody went: 'Wow!' (laughter) – and we went, 'oh I think the coup de théâtre works then'. And it was the best thing that no one had seen it.
>
> **Moran:** How much confidence did you have that it was going to work?
>
> **Mark Jonathan:** Well you know what, I don't know. What happened, I mean I had a feeling about something. In that piece I listened to the music – and I was hearing some lighting cues, that were about the pulsation of the music. And I looked at this gold back-cloth that Ruari Murchison had designed. And I thought, I'm going to put a circle of light on it and this is going to pulse with the music. Terribly simple.

> People saw the face of God – not just one person. Now the scenic painters had just gone splosh splosh splosh with some gold paint, and I'd just put this circle of light on it. And people saw the face of God – and people said: 'so you planned all that?' and you go, 'well – you know...' It went further than we'd ever intended. And that sort of discovery... I love finding things that you weren't expecting. I love suddenly discovering that something intended for Act 3 does something amazing in Act 1. You know, always check what you've got in your rig before you put another light up.
>
> But sometimes things are completely planned and work out. The thing is you shouldn't leave too much to chance. Let me put it another way: woe betide you if you haven't checked. Will the light get through the thing and under the what's-it and round the other thing? Or – where is that border in relation to the sight-line? If you haven't checked it, it screws up.

Jonathan makes it sound a bit like the response of the 'new star' being interviewed about his 'overnight success' – 'yes it's an overnight success built on 20 years of training, practice and hard work.' It takes a lot of something to make it look that easy working under the kind of time pressure exerted on lighting designers, particularly by some large and well-funded theatre dance and opera companies.

Limitations and constraints in rep

Dance productions, in common with most large-scale opera productions, are very often made for a repertory – that is, they play in the same theatre as other works, changing over as often as two or three times a week. This means the production needs to be technically re-mounted and the lighting design remade – usually without the original lighting designer present. Repertory production can also be 'put into store' and revived in subsequent seasons or slightly remade for different theatres. In our interview, Mark Jonathan commented that '*Sleeping Beauty* is scheduled to be in the repertory for 25 years.' We have already heard from Rick Fisher about going back to a production he lit ten years ago at the Royal Opera House, and not really feeling it was his. Michael Hulls talks about making work for a German and for a UK repertory, and ends up feeling like the work he made for English National Ballet was not really his, even at the premiere (and this despite the glowing reviews):

> **Michael Hulls:** Just recently, [mid 2014] we [Maliphant and Hulls] were making two pieces at roughly at the same time, for English National Ballet

and for a ballet company in Munich. The piece in Munich, rehearsals started early last autumn. You might have a week or two weeks, and then go back a month later for another week or two weeks, go back a month later for another week, then go back for a couple of days.

The premiere of the English National Ballet piece was two days before the premiere of the piece in Munich, so they were coming to the same premiere deadline pretty much. The English National Ballet piece was more or less one chunk of time. I think that probably the total amount of rehearsal time for both pieces was pretty much the same, but there was English National Ballet in one chunk and that kind of ends two weeks before the premiere, and then it is on. The other Munich piece was broken down over a longer period of time, and it was so much better to work in that way.

When it is just pretty much one continuous period, you always feel like there is not enough time anyway, because it is quite tight. So there is little time to sit with something and work out how to develop it, or say that this is proving a much richer seem to develop, concentrate on this, and let that bit go for a while. When you have got this kind of one chunk with a deadline at the end, and then you are out of time and then it's got to go on stage, that...

In those two examples it was such a clear example of how the process in Munich [short periods of working with the company and breaks in between] was much better. The piece was much better for it and it felt more like our piece than the English National Ballet piece, which felt much more like something for someone else.

Moran: The business of that cogitation, that development, that is allowing the artistic idea to develop, and also trying out technical solutions that enable that artistic idea?

Michael Hulls: Not trying out technical. That wasn't possible. ... [Y]ou had a day to light [each of] the pieces. But the Munich piece really was ... I think we had a day and then a couple of hours the next morning and then we did a dress rehearsal, 'Job done. I'm not coming back for the premiere. As long as you do just exactly the same again, I don't need to go back.' It felt right, it worked and did all the things that it needed to do. It looked good.

Whereas the English National Ballet piece always felt, 'No, I need more time'. And then also because Russell was changing things between the

> lighting time that we've had two weeks before and the premiere, then you've got, 'Okay, I can try and do this, this is the list of stuff that I've got and you've got that we need to fix. I've got 20 minutes and then we are doing dress rehearsal and that's it. Then I cannot change anything after that because it is just going to end up falling apart.' Those are situations where you have undertaken a commission for someone else.
>
> Say with *Shift*,[6] the piece with the shadows [that comes up in discussions in chapter 7]. Russell and I went off to Birmingham to the old DanceXchange, two days in the theatre with tape on the floor as a general principle, then moved into the studio next door. Put six brand new, beautiful Prelude 16/30s [small stage lanterns] on bench bases that I made, set them up, two days later, piece made, so in a very short amount of time. But then it's just Russell and me; he is dancing and I am doing lighting and – so it's not that limitations are not good.
>
> **Moran:** No.

There are always constraints, and the artist can often chose to use them to her or his advantage.

Abstract versus arbitrary in non-narrative

Given Paule Constable's frequently stated rejection of the arbitrary – something she is not alone in holding to – I wanted to ask Hulls for his take on arbitrary versus abstract:

> **Moran:** One of the things that Paule Constable has talked about, and she is an English Literature graduate and very much into starting by investigating the text, and I asked her whether she had done much in contemporary dance. She said, 'I don't want to do things that are just arbitrary.' But you're ...
>
> **Michael Hulls:** I don't want to do things that are just arbitrary.
>
> [In the work I do] it is hard to say what it is that isn't the text ... but it is something. I could say it is the sculptural form of a body moving, so it's like a kind of living sculpture that I have a non-arbitrary response to. I want to see it lit at that angle because then I really can see the sculptural nature of what is happening. For me that is not an arbitrary thing, that is aesthetic.

Moran: Do you see that in the rehearsal space? Do you see that movement, that posture in the rehearsal space, and know what are going to be the interesting angles to hit it (with light) from?

Michael Hulls: Either in the rehearsal space or from watching a video of rehearsal, or someone trying to work something out, so it is not quite formed.

... [I]t is a moment of inspiration and that can come from different sources, or something that you see and you go, 'that is such a strong image or visual...'

I was looking at a film, I think it was an old BBC documentary about Rodin, this was particularly to do with Russell's *Rodin Project*.[7] It was kind of research, and there was a shot of one of his studios with a row of half-completed figures and natural light in shafts through windows coming in. It was just like, 'Wow, that is fantastic.' There was something about the repetition of figures all being lit from the same angle but separated. They were in a line, it must have been skylights or something and the sun, and there were five different shafts of light through a window, all at the same angle onto these figures.

That helped me in the second part of it [*Rodin Project*], I want – wherever the dancers are, whatever they are doing, I just wanted to pursue always lighting it from the same angle. So 60 PAR cans or something, and it is all three-quarter backlight over the shoulder, which is one of the most sculpturally revealing angles to light something at.

Obviously it is a sculpture-based project; one of the things that Russell and I share is an interest in sculpture and making things as sculptural as possible.

Later the conversation contrasted dance's embrace of abstraction with the aesthetics and ideals contemporary drama. I mentioned Keats's 'Ode on a Grecian Urn' in this context, with its closing lines: 'Beauty is truth, truth beauty, — that is all / Ye know on earth, and all ye need to know.' Hulls sees himself as an artist working in light, interested in abstraction, but he does not reject the lure of beauty – something his work bears witness to:

Michael Hulls: I am interested in abstraction. I am not an intellectual. I want to be an artist and light is a medium through which I fulfil my own creative desire, as is making theatre. I am including dance as a theatre event. I used to do it in different ways, from making sets or acting or writing, and now light is the medium through which I engage creatively with the world. It is not painting or drawing, which it has been at other times before I got so involved with theatre. The fact is that it can be its own thing and that it does not really need explanation.

It's funny you mentioned the Keats thing, but I am unashamedly into beauty. There is enough shit and ugliness in the world. Not a superficial beauty; it has to have a beauty that emotionally connects with someone inside.

That is another thing that I am interested in; how the medium of light in contemporary dance, in conjunction with dance, [works]. In an abstract way that is quite hard to talk about, because we are abstracting it and we are allowing it to just be what it needs to be, but always with the intention that it has to connect emotionally with the audience, in whatever way that might be. And everyone will have a different perception and response to something, but light is psychotropic and it affects our mood. If it didn't connect on an emotional level then it would just be gloriously visual and beautiful, and kind of sterile and cold. I am largely talking in the context of where there is no other visual element than the clothes the dancers are wearing or not wearing, but there is no separate set design. ...

There is that fascination too with how something non-narrative creates narratives in people's heads, because we all find it really hard not to do that. Or how something so abstract and ostensibly meaningless can make a meaningful, emotional connection with someone. Sometimes it is hard to say how that happens and all one can say is that it does.

For me, Hulls' words here connect strongly to the idea I have attributed to R.G. Collingwood of the necessity for art to create emotional engagement. (Others have written with equal or greater eloquence on this thesis. Collingwood was the first approachable writer on the subject that I found.)

Hulls could be talking in this final paragraph about a painting by Mark Rothko as much as the abstract dance work he makes with Russell Maliphant. Just as some viewers of the painting will find an emotional connection to the Rothko through *seeing* a horizon in the gap between two colours, some watching Maliphant and Hulls' *Afterlight* will connect through the story the pieces make in their heads. Others in both cases will connect more directly to the abstraction, or to 'the things themselves' or through the things themselves to the emotion of the creators. Hulls' refusal of the arbitrary and instinctive embrace of abstraction help the work he makes with Maliphant, and others, to elicit an emotional response from audiences.

More generally, light for dance – relieved of the burdens it so often has on the drama stage of more direct and literal story-telling and of lighting speaking faces – is capable of connecting to and evoking emotion and connection in its own right. Of creating 'beauty that connects to someone inside' and of making meaning in and of itself, as an independent operator on the stage. As far as lighting design for live performance is concerned, dance does indeed lead the way.

7 In Search of the Right Light

Is lighting design art?

What grounds do we have for saying that theatre lighting design is ever art? Several text books on theatre lighting design have 'Art' in the title, but very few discuss the practice in those terms. With very few exceptions, theatre lighting design does not stand on its own but as a contribution to a larger whole, the production. It requires reproduction which immediately raises questions when it is compared to traditional art forms such as painting, and it cannot be put in a gallery in the same way that a Rembrandt self-portrait or a Duchamp found object can be. It requires collaboration. The technical collaboration of, for example, lighting crew, programmers and stage managers, can be compared to that of the skilled artisans working alongside Michelangelo, Rodin, Jeff Koons or any number of recognised fine artists from almost any period. However, the collaborations with directors, choreographers set and costume designers, and with performers too, more closely resemble collaborations on a film set or a fine-art photo-shoot, and raise more questions about ultimate authorship. Whilst copyright and contract law acknowledge the authorship rights of theatre lighting designers in the UK and elsewhere, many theatre practitioners, cultural commentators and critics attribute authorship of a theatre production solely to a director (or sometimes a star performer). So what do lighting designers themselves think?

> **Moran:** How comfortable are you with the idea of the lighting designer as artist and with lighting design as art?
>
> **Mark Henderson:** I think it can be ... Again it is a collaboration isn't it? Because you can't generally do it on your own. So you are using other people, and other circumstances – so I guess it's not purely your piece of work. It is an art and it is artistic, but not purely on its own.

Bruno Poet is even less sure:

> **Bruno Poet:** I don't know what I am. I like turning lights on and off. I sometimes describe lighting as an art for people who can't draw. ... I like making pictures. I get pleasure from making pretty pictures on stage. Not even pretty pictures – but nicely shaped pictures.

When I began talking to Neil Austin about this, his first point of comparison was Old Master and Renaissance painting. I proposed a broader definition of art and artist, having just re-read social philosopher Raymond Williams' *Keywords* (Williams, 1976). Even at the end of this conversation I was not sure if Austin did or did not see himself as an artist:

> **Neil Austin:** [Theatre] is a collaborative art form and there are many different pressures. Let's say there is a commissioner – the producer or the producing house – and yes we are being asked to tell a story. Like them [the Renaissance painters] being asked to tell one of those biblical tales – there's a similarity there. Are we artists? I think we are probably very skilled craftspeople actually. I think to say that we are artists is pretension.
>
> **Moran:** I tend to be on the other side and think you are over-valuing the word 'artist', that's all.
>
> **Neil Austin:** Yes but I think that's where the pretension comes in. The pretension is wanting to aspire to a name and a word that suggest that you belong alongside some of those names that would be indisputably, in the public's mind, artists.
>
> **Moran:** But there's a lot of minor artists that the public would still call artists but whose level of talent and impact on the world is considerably less than ours.
>
> **Neil Austin:** OK. In pure semantic terms you would call a great dancer or maybe a great actor an artist. So maybe in those terms, you could. I would suggest there is an awful lot of craft.
>
> We are very, very, collaborative artists. It is an interesting version of art where many different disciplines come together to create a single work.

Other lighting designers are much less equivocal. Here is Katharine Williams and then Natasha Chivers:

> **Moran:** Do you consider yourself an artist?
>
> **Katharine Williams:** Yes of course. I think it's a bit of an anachronism for someone to say that we're not now. And like you said, I get that there are people [doing the job] who are not. But yes of course we are.
>
> **Moran:** Where do you sit with 'is lighting design art?'
>
> **Natasha Chivers:** Definitely it is. It is, unquestionable. If you look at a painting or you see some set-up for a high-end photo shoot – and the time it takes, the complexity and the composition, all for that one still image of a great painting or photograph – and we do that time and time again, we may do it 200 times for an evening. And not only composing the pictures, but the decisions about the pace and the focus for each moment that go along with that. I think every solid bit of lighting design is an outstanding achievement, and that's why there aren't that many people out there that can actually do that. If you think about it in those terms – that we are creating and composing constantly shifting stage pictures – thinking about the pieces that you're most proud of, there's a high-impact photographic image that's been composed, and then there's the movement between those images – how can anybody question that it's an art?

Although Johanna Town's training and much of her professional background is in technical roles (she became the chief electrician at the Royal Court theatre in 1990 and left as head of lighting in 2006), she is equally emphatic:

> **Johanna Town:** I am an artist. I don't think it's the answer people expect from me because I spent so much of my life coming up through the technical field.
>
> One of the reasons I went into lighting was because I couldn't coordinate a paintbrush to a piece of paper, and yet inside me I had all this pent up energy I wanted to get out, and lighting was a way that I found when I was in my teens that I could create emotions and feelings and shapes and forms that I couldn't get onto a piece of paper with a paintbrush.
>
> I continue to work in that way; it all has to come from the soul. And to me that's all it has to be. I wouldn't say that the work I put on stage is always like a work of art.

> Occasionally that happens because the set and the show and everything allows it to happen, but just because it's a one lighting state, living room comedy drama doesn't mean that it hasn't come from some part of your soul to arrive on that stage.
>
> Nobody but me would feel that my soul was in it, I don't think. I think there are people in the industry who would go, 'That was beautifully done, that one state or that living room drama,' but it doesn't necessarily mean the audience would think it was a work of art or a piece of art.
>
> I know it's come from inside of my soul rather than just [me] doing a job.

Jon Clark trained at Bretton Hall, on a course that was soon to be assimilated into the University of Leeds. His was not a technical training, though he did a lot of technical work in his early career. The course clearly helped him to see his practice as that of a creative artist, as he explains here:

> **Jon Clark:** [At Bretton Hall] we were not taught to be a [particular kind of] practitioner. It was more about making theatre. For us it was a designer's performance a lot of the time. So, of our major assessments, one was about making individual designs for a performance-led piece, without performers in it, and another was a group piece on that level too. It wasn't about assuming any one role. It was about making theatre from a design or spatial point of view rather than learning how to perform a specific role.

It's not all art

Not all theatre lighting design is art, however. At the end of chapter 1 Paule Constable said, 'Lighting can be very pedestrian,' while James Farncombe is not alone in admitting he is not always able to make the kind of response from his soul that Johanna Town talks about. The following is an extract from an email Farncombe sent me after the interview:

> I think whether or not a lighting design is art must be to do with intention, context and, to some degree, integrity. The lighting on a station concourse, for example, might be installed with the simple intention of providing illumination. In this context it fulfils nothing more than an intended practical purpose and therefore it has no artistic merit.
>
> But the exact same lighting might be installed in a context where the intention is to evoke a response from a viewer or viewers. The light might be employed to elicit associations with emotion or memory. Certain

conditions might give it metaphorical significance. In this context, it may be said to have artistic merit.

So if, in the context of a theatre performance, lighting is employed to elicit a psychological or emotional response from the viewer, and in doing so it alters or develops the viewer's experience or perception of the piece, the lighting has artistic merit. In other words, if the piece as a whole is diminished or even rendered meaningless without that [particular] light, then the lighting has artistic value.

I would argue that lighting design can be rudimentary and still be of great artistic value if the intent is carefully considered, creates a strong aesthetic and clearly communicates information to a viewer. Equally, a hugely complex and technically brilliant design might be considered little more than spectacular decoration if it serves no purpose other than to stimulate the eye, like a firework display.

Who cares?

To close this section, who better than Ben Ormerod:

Moran: Is there a justification for theatre lighting design being called art, and for theatre lighting designers being artists?

Ben Ormerod: Okay, so my answer to that is, 'Who cares?'. That's the first response.

Secondly, I do think that theatre is an art form, so ... there's an interesting question here, isn't there, which is to do with what's the difference between a lighting designer and an electrician? That, in the end, is the question, and this is the only answer I'm going to give you, as to whether or not a lighting designer is an artist. It doesn't matter what the show is – if you changed the electrics team half way through the tech of the show, the work of art that we call 'a show' that resulted at the end would be the same. If you changed the lighting designer halfway through the tech, that work of art would be substantially different. That is my answer to that question.

... The piece of work that I'm most proud of is the Calico Museum of Textiles in Ahmedabad,[1] which I've been lighting for about 18 years and I'm going back to in October to do some more work. There's nothing of me in that, it's a technical solution to a particular, unique set of problems. What

> is in there is my experience and my love of materials, but it's not me. I'm sure, if somebody else had done it, it might have been done differently, but that doesn't mean that there's anything of me in there, I know that sounds paradoxical.
>
> On the other hand, if you go and see *Tristan*[2] next year and I do a good job, whatever that means, there will be more of me there on stage than there is in front of you now.

Practice and developing an instinct

Throughout the interviews, lighting designers have talked about 'instinct' informing their technical and aesthetic judgement of what is the right light. Instinct implies something innate, yet the lighting designers also frequently talk about *learning* or *developing* their instincts. What can be said is that none of the LDs here claims to have been able to consistently do artistically interesting work from the beginning of their careers.

Natasha Chivers talked about learning from tutors at LAMDA (where she took the two-year Diploma in Stage Management) from the peers she worked alongside early in her career, and from reproducing the work of other lighting designers on tour (re-lighting):

> **Natasha Chivers:** One of our acting tutors at LAMDA set up a studio theatre in a basement, basically for people from the course who wanted a bit more development or an alternative to going for a 'rep job'. So we raised the money, built the seating rostra, put in a grid, a box office and a café that we ran. So I did a lot of lighting there – but I was really stubborn and I thought I knew it all. ... From there I met a director and a set designer – Rufus Norris and Katrina Lindsey (who you may have heard of) and we set up a company together for about six years. They were a bit older than me, so they kind of taught me (without knowing it) how to be part of a creative team really, including what terminology to use. I wasn't used to flexing that creative muscle, so that was great.
>
> Other parts of it I learned from doing it, creating my own lighting designs in small places, climbing a ladder and putting up my lights myself and gaining that kind of confidence. Trying out design ideas and failing and trying again and getting it a bit more right.

> And then from reproducing other people's lighting designs [re-lighting] on tour. That helped my confidence as well. [What was great about re-lighting was] getting hold of someone else's lighting design, where the lion's share of the creative work has been done. The design works. [You then] digest that, and reproduce it yourself, still having a major investment in it but not having total responsibility for it. That was really great.

Chivers went on to talk about how her instincts developed further as she built her own independent practice:

> **Natasha Chivers:** I remember there being a point where – well the stages are: the blind rabbit in the headlights, and there is a good deal of panic in that, and then the stage when you've got to know a bit more about what you're doing and every job is a massive undertaking, and you're learning big things all the time, and slowly you're able to build – as you're doing something not for the first time – so there are only so many ways of doing windows, and only so many sorts of windows that you're going to get – and you'll light wooden floorboards for the first time, you have to learn what colours work best.
>
> So then I remember getting through that phase, and I'd done everything at least once, and it felt like the adrenalin and nerves and all that sort of thing had settled down. I would continue to challenge myself by using colours I didn't know to get that [the adrenalin, etc.] back, but then you realise that you're never going to get that thing back in quite the same way, but that there are still going to be challenges and you're still going to have doubts.
>
> And then you get to the point where other bits of your life are exciting and maybe you don't feel the need to get all your adrenalin kicks from the work. But what's really lovely about where I am now – and we've talked a little bit about this at the ALD and elsewhere, and it's the stuff that nobody really tells you about. Stress management. And you come to the realisation that being stressed is very exhausting – and so being able to manage the stress of the work – because you have a much better understanding of how to do the work – releases energy that would otherwise be consumed by the stress. So that energy now is free for the detail of the work. ... It's incredibly satisfying.

Michael Hulls on practice and experiment

For Chivers and many of her peers, experiment happens on productions, in front of lots of people in a tech. That can be fine and, as Chivers insists, it requires trust. But it can also be both intimidating and limiting. Michael Hulls emphasises his need to experiment, and through his long-standing collaborations he has been able to do that away from the pressure to produce quick results. His wish to have time to think about how an idea might work and develop echoes something James Farncombe said in interview, dreaming of being like Michelangelo lying on his back for a whole week looking at the sky!

> **Moran:** Is it [learning how light works on stage] like an artist in a way learning how to use a pencil or a paintbrush? How long do you think it took to sufficiently master those things, and where do you go to keep developing that?
>
> **Michael Hulls:** Trial and error, through experience of doing things.
>
> I guess what is difficult about lighting design is, to use your analogy of drawing or painting, to master those media, you practice. If you are a violinist or a pianist, you practice. You practice for hours and hours and hours. Unless you have got your own lighting studio, you cannot do that. So [for me] it happens through having R&D periods where you can work with light without the pressure of it having to produce something. Through having sufficient time to work at something and keep coming back to it, and keep changing it and refining it, like a sculptor chipping away at something; you're still not satisfied and you're still looking for something. You just keep going at it; obviously time and money are massive restrictions on that.
>
> For me, those are where it happens; through R&D time or through the lighting time that you have on a piece. When Russell and I are working together, our ideal is that we start with lights, be in a space with lights on day one, and then get enough information about something to then be able to develop it with dancers in a studio; for me to develop it in my head or through plans and scale drawings of angles ...
>
> Then we go back into the light halfway through our creative process, have another time where [we] can then see how things have developed. Maybe they are working out or they are not, or maybe then it goes in a slightly different direction.

> ... The older and more experienced I get, and having different experiences, that's the kind of ideal creative process, broken into periods with gaps in between where you can look at the video or think about it, and then work out how to develop it ... I think that is important.

Lucy Carter asks, can lighting design be taught?

Lucy Carter, who as we have heard also enjoys experimenting away from the pressure of the large-scale theatre tech (see chapter 6) acknowledges that her practice is built on her years of experience, but...

> **Lucy Carter:** The question that I still have is can you learn to be a lighting designer or do you have to have an instinct for it? I think you absolutely need to learn the skills: you have to learn how to draw a plan and how to analyse things, and talk to directors, and programme and all of those things you absolutely have to learn ... I did one year [of formal training, an MA at Central School of Speech & Drama] and certainly wasn't ready to go out and just be a lighting designer. I still needed to learn the business. What it's like to be out in the professional business of working in the theatre? You don't have to do it as a technician, you could do it as an assistant, you could do it as a [programmer]. It's really hard to say but there's definitely instinct but then there's the craft.
>
> **Moran:** Andy Bridge [who lit *Phantom of the Opera* and a lot more] used to talk about the split between raw talent, technical knowledge and people skills.
>
> **Lucy Carter:** I have to say a massive part of my job is people skills and this many years in to my career, I feel like 'my God I manage all these people!' I'm not their boss but I have to motivate them, get them to want to do it, inspire them to help and make a good team to get the finished product ... without any management training whatsoever!
>
> All my friends who work in business have done so many courses in management and I've had none! You learn it by the seat of your pants, but you still fail miserably sometimes. How to get that grumpy teenage follow-spot op to care about what they're doing for you, and not feel that they're just being told what to do. I don't know the answer to that, I really don't. ...
>
> ... I think at this point, 20 plus years in, realising that this is my career – because up until now I've been doing it because I love it and I've been lucky enough to keep being offered work in it and I keep doing it and

I'm earning a living from it – but now at this point I realise, 'oh yeah, this is what I do. It's not just by chance anymore.' ... Realising that you think in light and cue structures. I can't really explain that but I'm constantly thinking, 'Oh that would be great lit up like that.'

Time to finish the work

Rarely in the UK will you hear a theatre lighting designer saying they have had plenty of time to do their work:

> **Johanna Town:** I do wish we had a lot more time, like the Germans, to actually create our art.
>
> **Neil Austin:** [E]ven the best of imaginations can't foresee exactly what it's going to look like until you are in the space. That's why lighting time is very important.

Even with 'the best of imaginations,' the kind that comes from long experience and great instincts, the *right light* these lighting designers are aiming for can often only be seen, judged and revised in the context of a performance, or at least a full dress rehearsal. Here Mark Jonathan responds to my question, 'how do you know when you have finished?'

> **Mark Jonathan:** I always used to say, 'You know when it's finished because the show's open and you're leaving to go and do the next job.' Actually, what I've started to do with ballet – because that seems to have the most intense schedules and the least amount of time and directors or choreographers who say, 'That's great but you know that bit when – can we just do a bit more here or a bit more there?' Because they've been so busy choreographing the ballet, they're only now looking at the picture on the opening performance. I often say to ballet companies, 'I want to stay another day or two. You don't have to pay me any more money but I need the set and a lighting programmer to just finish things off.'
>
> I might just go in later the next day; it depends if I'm overseas. It is better to see more than one performance as the piece beds in. At least we can just carry on tweaking a little bit, finishing things off that were rushed, especially when the production is going to have a long life. I've gone back to a show two years on from its original premiere at the Royal Opera House to try and do the bit we never quite got right the time before.

Moran: Yes, how do they take to that?

Mark Jonathan: It's okay. I say, 'I need a little bit of time just to do that.' They'll go, 'Not today; we've got the changeover to *Turandot* today. Maybe tomorrow.' Then, tomorrow you say, 'Shall I do it?' 'No, no, no, it's another big opera tonight. Maybe the next day.' Then, finally, you say – this really happened – 'You know the transformation scene? I haven't really worked on it. I just need a bit of time', meaning 60 minutes. The supervisor said to me, 'I'll give you six minutes.' You consider momentarily arguing with him for five of those six minutes and you realise you're only getting six minutes so you do what you can.

Often, I'll go look at my lighting when the piece comes back in rep and I'll go, 'Why does a cue look like this?' This is the truth; it looks like that because I've placed the cues in the score where I think they're going beforehand. Then, in the rehearsal the stage manager gives the next cue and I say, 'Okay, record', and I start working on the next cue. Depending on how much time there was between one cue and the next is how much time I've had to hone that cue before the next cue comes. At the Royal Opera House, you sometimes get ten minutes to light Act 1 before the 11 o'clock rehearsal. You get ten minutes to light Act 3 after the end of the rehearsal. Act 2 you had better get right over the rehearsal or you do it on a rare 'technical' Sunday.

So your question is, 'How do you know when it's done?' My answer is, 'I did what I could in the time.' It may not be done, actually.

... Of course, it can be a bit like, 'Don't compare yourself to Michelangelo, Mark Jonathan.' (Laughter) You know how he paints himself in the wall of the Sistine Chapel? At the end ... the day of judgement. His anguished face in there because he's painted the ceiling above it. It's hard for a lighting designer because, finally, on the opening night – I sometimes can't sit in the auditorium anymore – you sit there without the headsets on. You hear the play or the music again without all that babble that's going on over the headsets. ... You start to sit there and go, 'That's not very good' or 'He's not lit there' or 'I should've done a bit more there.' You have to be able to say, 'I did what I could in the time.'

Jonathan is expressing some frustration here not only with a lack of time to make the work, but also hinting at what can sometimes be another frustration – the 'babble' in the headsets that can get in the way of creativity, of making *art* on stage.

Previews, refining and polishing

Johanna Town uses previews to take notes and refine the design. She finds that it is often useful to experience the piece in front of an audience in order to complete her job, to find the *right light*:

> **Johanna Town:** I hope that every show I do has an impact on the audience. I may not always achieve it but that is my aim. I also understand how not being able to see a show or not understanding the shape or format of a show through the lighting can completely ruin a show for an audience.
>
> I quite like lots of previews because I like to be able to sit all over the place. It's really interesting to see when you move away from that production desk and realise the audience isn't engaging because it's probably under-lit. I think you can have a real sense in the audience of your impact on the show, and I think that's really important. …
>
> That whole thing of sculpturing – the re-invention you might have in the tech, the naturalism of it all – you get all of that there at the first preview. And then you've either got it right at that point, or you have to lift it to reach the point where you've got all those things *plus* what the audience need, which is to engage in it.
>
> It's very easy to sit and watch a show and know every word that's said because you know it so well. You've got to make sure that everybody else can hear and see what's happening. That is very, very important.
>
> I've tech'd cues again and again and again trying to get the boomph to happen at the right point or whatever. You're sitting there in your post-dress tech sessions fine tuning with the actors, and they might do it 20 times, and then you sit in the first preview and you go, 'That's where it goes'. Suddenly you're away from all the machinery that makes it happen and you're sitting back being a member of the audience and you know exactly where that cue has to go when you see it as a whole. You're emotionally watching the piece and not hiding behind [the machine].

It is interesting that Town talks about the production desk – her workplace during the technical rehearsal period – as 'the machine'. Interesting, too, that in order to achieve the naive viewpoint that Ormerod mentions in chapter 5 (seeing the piece as if for the first time) Town has found being in the audience useful, away from this *machine*. In some ways this is similar to Jonathan's relief at being away from 'the babble'.

Key here, though, is Town's desire to create affect, and belief that when lighting is not *right*, it can fundamentally undermine the communication between stage and audience. Like Town, Paule Constable freely admits that she does not always get it *right* for the first preview. She talks about needing to polish the work, a similar analogy to that used by Farncombe (who compares refining his design to creating 'sheen and luminosity' on a piece of wood, in chapter 4). She also mentions instinct:

> **Moran:** when you are polishing, through previews, for example...
>
> **Paule Constable:** I think it's an interesting mix of intuition and instinct, being faced with what's in front of you. I think the real test through previews is about the aliveness of it, the story-telling of it, the front-footedness of it, the way it really works. You can have it in the abstract or you can see it episodically, but ultimately – what does it feel like to be in the room with it. That's what you're testing there. Things like the rhythm of change and that sort of thing, that you really look at in previews. ... Do you feel like you're in the room or do you feel that you're being denied? Are you being taken somewhere? Does it jar? Does it work? Does it land? What can you do to share it (the best bit of the experience of being in that room) better?
>
> **Moran:** And with a piece like *Wolf Hall*,[3] that started on a thrust stage and moved to a proscenium stage, does the intention remain the same even though the way it is achieved technically may be different?
>
> **Paule Constable:** Yes, I think it does. And it is interesting with something like *Wolf Hall* because the rules became so very different. But the aesthetic, the sense of how light energised people in that space, was the same. The big key dynamics, and then feeding in underneath that. The sense of greyness, the sense of darkness, I think is the same. The rules of how it shifted became very different.
>
> I still think the best way you learn is by working in the round, because you can't step out – you've got to stay inside, you've got to work from inside, and it's really hard. But it's incredibly good for you. ... That thing about the figure in space, and the deal you make with the audience about that, and it's constantly changing, you can't constantly serve them up in the same way. I love all that.

Affect, impact and a little bit of phenomenology

In putting into words what Constable is looking for – this combination of 'intuition and instinct' – she uses a broad range of analogies to describe that which may be beyond description – the unique experience of her work as art. Not only does this touch on reason three for why lighting design is under-represented in academic analysis of theatre, it perhaps also suggests that a phenomenological approach to analysing lighting may be fruitful.

Phenomenology celebrates 'the things themselves' as potential carriers of meaning. Certainly for Michael Hulls, comfortable as he is with abstraction, his light on stage can be 'about the thing itself'. His work often has at its centre the idea that light is its own independent thing on stage. I am arguing that others here do this too, though they may be less comfortable admitting that or even recognising it themselves, for a variety of reasons.

Hulls has an understandable professional interest in perceptual phenomena:

> **Michael Hulls:** An interviewer yesterday asked me if I thought there were rules, and were rules a good thing? I said that generally I am of the opinion that rules are a bad thing and are there to be broken, but I don't really think there are rules. What I think there are, are perceptual phenomena – the way our eyes work. They are not rules. They are things like – the eye will always be drawn to the brightest thing. That's not a rule. That is just the way our eyes work. So you need to consider the way that we respond to light both on a perceptual and on a physiological kind of level.
>
> **Moran:** Such as the particular lighting conditions where fast-moving limbs begin to have an after-image ... that you've made use of, yes?
>
> **Michael Hulls:** Yes, and that is something I am really interested in. *Two,*[4] which Sylvie Guillem has danced a lot, is a piece where we (Maliphant and Hulls) were interested in looking at that. Some people said, 'How do you do that' [make Guillem's arms appear to multiply by creating strong after-images]? We said, 'Well, we've got this special solution called "Arms of Fire" and she just paints it on'. (laughter)
>
> What is really interesting is that the lighting is being made in between your eye and your brain.

This realisation is important for everyone working in theatre. This is perhaps the most important difference between the way a camera *sees* the stage and the way an audience experience the stage image. Humans fundamentally lack the objectivity of cameras. There is always interpretation involved in human visual perception.

Hulls went on to explain how he sees the difference between the way he works with Jonathan Burrows and with Russell Maliphant, and how this reflects the differences between the ways their works are experienced by audiences:

> **Michael Hulls:** I think that when working on something like *Stop Quartet* or other work that I did with Jonathan, which were more to do with pattern and rhythm and structure – and this is probably a load of codswallop, but it is my codswallop so I like it – I would say that I was experiencing those pieces somewhere near the front of my brain. And that when I am working with Russell, I am experiencing the pieces somewhere closer to the back of my brain. Now we might say the conscious and the subconscious, for example, rather than ... front and back. So I think that we experience different types of dance work or lighting in different ways or with different parts of our brain.
>
> A neurological map of which parts of the brain are working when seeing different things would either prove this is complete codswallop or whatever, but that is how I sense it, or characterise it.

Any interested brain researchers please take note. Hulls continues, and in the process shares more interesting results of his life-long practical study of human perception:

> Jonathan's work is much more brightly lit and Russell's work is much more ambiguous. There is something about the [low] level of light [in much of my work with Russell] – what they are doing is slightly ambiguous because it is not completely exposed. You are highlighting some things and lessening the emphasis or hiding some other things.
>
> Sometimes it is important to hide things with light; it was important to hide Daniel's feet in *Afterlight* because what is going on with Daniel is all upper body. What the legs are doing is not important. It is all these angles that are important.
>
> There is something about a low level of light and an ambiguity where you are not quite sure what it is, that I think increases your sensory awareness, that you hear more. I think it is that fight or flight response because you are not quite sure what it is. There is something moving in the shadow

there, and because you are not quite sure what it is, all your senses open up to work at a peak function to try and work out, 'Do I like it or do I not?'

Moran: That's very interesting. During *Still/Current,* I was watching what [the audience] were doing, and there were moments when I think I saw that happening, people behaving anxiously, moving forwards in their seats, and then there were moments when they were more relaxed, and sitting back.

Michael Hulls: Mmm... But I have that sense of, Jonathan is at the front of my brain, Russell it's more at the back. With Jonathan it's that things need to be clear because they are so complex, whereas with Russell it is moodier and there is an ambiguity.

The neurological accuracy of Hulls' analogy is not the point here. Neither he or I are brain scientists. The point is he feels that there are distinct differences in the way he reacts to the source material – a more cerebral reaction to the highly structured work of Jonathan Burrows and a more visceral reaction to Maliphant's work.

Also worth noting here is his conscious use of low light levels to provoke primal anxiety in the audience. I'm no evolutionary biologist either, but this seems to be a useful insight.

Hulls then talked about a particular response to *Shift,*[5] a solo piece he co-created with Russell Maliphant, in which Maliphant dances with his own shadows. At times the shadows appear to be dancing with each other. In the piece, six lights on the floor at the front of the stage cast shadows of Maliphant onto screens upstage. As experienced lighting people, we would both assume the audience would quickly work that out that each of the shadows is of Maliphant, but clearly not all of them do:

Michael Hulls: In *Shift,* Russell's piece with the shadows, after a performance once somewhere in France maybe, this woman came scowling up to me at the lighting desk at the back of the auditorium and said, 'Why didn't you let the dancers behind the screen come out and take a bow?' And I said, 'Because some of them are really big and ugly and we don't want to frighten the audience'. She kind of scowled at me again and walked off.

I thought, 'Musician. This woman inhabits a completely different universe to me.' I don't understand the universe that is music, and the way that it operates, it is kind of beyond me. It is a foreign universe that I am aware of but cannot understand. I could say something equally weird about music because I am from a different universe; I am from a visual universe. I thought, 'God, the world must be full of the most amazing phenomena to other people'.

The lights are on the floor, and that was in a place where there is the stage and the audience are raked and the lights are right in the front.

Moran: So they could see the sources.

Michael Hulls: You could-

Moran: But people don't, [put two and two together] do they?

An example of the magic of theatre perhaps? In any case it underlines Hulls' point that we see things differently and reminds all visual artists that they don't control the meaning their audience take from a piece.

Peter Mumford, too, has worked extensively in contemporary dance, and, like Hulls, uses these opportunities to make light on stage be *the thing itself* (for example in *E=mc²* mentioned earlier). Like others here, he believes what he does with light on the drama stage creates its own independent beauty:

Peter Mumford: I think I am trying to have an impact on the audience, in whatever I'm doing. You could say that something like *Ghosts*[6] is a very straight play in a sense. But light, certainly in that production, plays a major role. Not to overwhelm it. That's the thing: generally speaking, you're not trying to impact the audience by giving them some kind of spectacular light show thing. You're trying to impact the audience via all the other media that are at work within the piece. And you are. I am trying to do something that is beautiful to look at, whether that's lighting a person or a step or whatever. And I'm very aware that I want the audience to like that – I want them to enjoy it.

Back in chapter 1 Nick Richings said, 'It would be a very dull play if you just had grey daylight streaming in all the time'. Strict imitation of naturalism – if it is even possible – is not usually interesting, it's not *dramatic*. Yet at the same time the drama lighting designer is often aiming to help evoke in the minds of the audience a specific time and place on stage. The challenge is to create light that is affecting and evoking at the same time. Bruno Poet here talks about this in relation to rendering night time on stage:

Bruno Poet: *Timon of Athens*,[7] which I did with Nick Hytner at the National was set sort of now – in the present day – and we had various

> scenes where we liked images we'd seen on the television. So I downloaded a few news references – riot police and the riots in London[8] and that, just to get a sense of street lights and colours. It's very easy in theatre to do sort of, 'it's night time – it's got to be blue moonlight', but I wanted to look at it and think again... So just downloading some images, or just going out and taking some photos of things we see every day, and thinking, 'OK – that's what it really looks like'. It's funny how we often don't really look at what's around us.
>
> And then there's the whole theatrical language. You can faithfully recreate a city at night, and then you go, 'well why is that red? That's not right. It's not that type of scene.' And you're telling everyone its night time but maybe the red neon sign that you're imagining just round the corner doesn't actually feel right here. So you know you've got to balance it, but I think it is very useful [to start from a picture of reality]. Rather than to try and start from scratch, you start from that sort of thing.

In chapter 3 we first head from Natasha Chivers an idea that the lighting designer is more often performing a role like that of a rhythm player (rather than a lead instrumentalist). We have heard, too, that the *right light* is not just about the cue state, but also about the transitions – technically, the times recorded as part of the cue state and the placement of the cue point and aesthetically the way the light moves in space and time on the stage. As Lucy Carter said in chapter 6, 'we all know, if you get into a cue in the wrong way, the cue can just feel wrong, even if it's right'. This all speaks to the need for the lighting designer to be – at the very least – sensitive to the underlying rhythms of the piece, and often to be subtly driving those rhythms, not unlike a good jazz drummer or bass player. Here's Bruno Poet again:

> **Bruno Poet:** I think that [theatre] audiences see lighting in a very subliminal way. I don't think they are necessarily aware of what's going on – which is fine. So you have to be working in a subliminal way. I guess it is part of the [lighting designer's] instinct because if you get the transitions to feel right, and they are driving the piece, and they are paced right, then it is affecting the rhythm of [the whole piece].

The right light? (or the route to a right light?)

I have argued from the start of the book that the 'rightness' thing is not in any way to say, 'There is one perfect solution, how do you find it?' It is to say, 'You have found a solution, how did you come to that solution? What were the routes that took you to that solution?'

It has not been a simple question to answer:

> **Moran:** So the overarching question is, 'what makes the light on stage for any particular moment in a production the right light?'
>
> **Mark Henderson:** Yes – that's a difficult one because everything impacts on everything else.

In chapter 1 Mumford says that modern audiences' familiarity with cinema and television enables them to, 'read a language of editing ... of parallel action' and a lot more, 'without even knowing it.' This mostly subconscious knowledge could be described as visual literacy.[9] Here he expands on the implications of these subconscious skills and suggests that today's audiences have a different level of visual expectation from earlier generations too:

> **Peter Mumford:** I find that audiences are getting more and more aware of what's going on light-wise – not that they necessarily understand how it's achieved – but they are becoming much more aware of the visual, and also I think more demanding. I think audiences expect a certain standard, a certain visual excitement from what they are looking at. I think they expect it to do that, and to look good. For plays to have a kind of painterly finish or whatever. Obviously in a big brassy musical you've got the opportunity to do something that's a bit flash. But what's interesting to me is when you're not being flash in that kind of a way but you are creating something that's very beautiful to look at.

With Paule Constable, the discussion of the need to balance potentially competing needs while making something that we can describe as beautiful, referenced various ideas about the nature of art, including Immanuel Kant's (1724–1804) notion of purposefulness. This is not an attempt at a philosophical discussion, and we don't discuss any of the well-known problems with Kant's theory, but instead use it as a shorthand in an attempt to be clearer about what we mean by the 'rightness' of light for a particular moment on stage:

> **Paule Constable:** So in a way, it is by... taking everything back to its simplest thing – this is about light through a window, or this is about this or that light, then when you add, you become very aware of what the repercussions of adding are. So it is trying to find ways to support the text when its needed, but you still maintain a live space. So you can still maintain something beautiful, really. Of course that's a subjective thing as well, isn't it?
>
> **Moran:** Kant talks about purposefulness, by which I think he means the rose is beautiful in part because it is very good at being a rose.

> **Paule Constable:** Yes, and that's like great design as well isn't it. It looks extraordinary and feels right. Ergonomics. Function and form. It is all those things. I think that's very valid – the idea that something is purposeful.
>
> **Moran:** So I'm thinking the light has its purpose and its beauty – and its quality as art is depends on both?
>
> **Paule Constable:** Yes, and that's like going back to all of those things [we just talked about] for example: 'Do you feel like you're in the room or do you feel that you're being denied? Are you being taken somewhere? Does it jar? Does it work?'
>
> When they all land, that's when it gets good.

One key quality that we seem to be looking for, then, could be called purposefulness. The same quality the rose has; aesthetic beauty of form that also fulfils function – a beautiful rose is also very good at being a rose and a beautiful lighting look is also very good at doing the other things it needs to do for the production (support the text, reveal what needs to be revealed in the right way, etc.). An artistic lighting designer, working in an artistic way, has to be able to plan for this, and has to be able to spot it if it happens by happy accident.

It seems to me that what is important is to recognise when it is working, when the technical and aesthetic choices are appropriate in that moment and for the whole production.

> **James Farncombe:** You can be objective about it. You stand outside of it and you see it for what it is and you suddenly think, 'The whole is greater than the sum of its parts.' The way that you got there isn't always clear, and a lot of luck is involved, and sometimes, at those points, you think, 'that's a piece of art. That's speaking in a way that's taken on a life of its own.'…
>
> The process of being able to stand outside it and recognise it [as *right*] is as important as planning it, knowing when something is *right*.

Light is the water that the production swims in

After all the interviews, after reading them and re-reading them, and after years of my own practice both as a lighting designer and as a teacher, this is the analogy I've found most useful. Stretching Jean Rosenthal's analogy beyond dance, light is the water that the production swims in. Sometimes it is still and transparent, hardly noticed at all. Sometimes it full of darkness that creates perceptions of threat and foreboding. Sometimes the production glides easily in a steady current, or moves rhythmically in gentle waves. Sometimes it swims aggressively against the current. Sometimes the water is restful and soothing, sometimes it threatens to disrupt, overwhelm or worse.

Light on stage rarely works alone, and it is part of the thing it supports – this *water* is inside the production as well as all around it. Its form and its function come from within the production. Its curves originate from the curves of the production, though they may be transformed or distorted to support an *active aesthetic*.

When it is right it enables the audience to experience the same *water* that the production is moving through, to taste and smell the essence of the production in the foam and spray of its breaking waves perhaps. It supports a connection between stage and audience that allows meaning to be made by that audience as individuals and to some extent as a group too. The quality of the connection may be clear or scrambled or anything in between, but when the light is *right* the connection will be right for that moment.

When it is *right* it is purposeful – it has a *right* aesthetic form and fulfils a *right* function for that moment of performance.

When it is *right* for the whole performance, it is clear that all this water is the same river, lake or sea. It has a unity that runs from the first to the last moment of the performance. If the production needs a seamless flow from one moment to the next it changes pace or depth, darkness or clarity, appropriately. If it needs to draw attention to itself, to tell its own tale or to distract, it provides cataracts, waterfalls, whirlpools or even a tsunami.

And at times, it can support the weight of meaning alone. Light on stage can be the thing itself. Abstract or figurative signifier, it can tell its own stories and celebrate its own form, its own wetness if you like.

The right light, the right LD?

So what is the *rightness* we celebrate when we see artistic light on stage? We may propose a set of conditions, none individually sufficient on its own or necessarily always present:

- Almost certainly it has involved collaboration, with technical *and* creative teams, which has to a greater or lesser extent been managed by the lighting designer as an integrative practitioner.

- It has an active aesthetic deriving from the production. The choices made in its creation 'may not be complicated, however there is a certain level of complexity involved in the planning' (Bryon, 2014, p. 193).
- It is made with a real understanding of the piece, and from a point of view that is in harmony with other members of the creative team, including key performers. This understanding may have been reached through academic analysis or through a mix of instinct and intuition, or most likely all three.
- It does not get in the way of the story-telling. It reveals what it should, when it should and how it should – including, when appropriate, the faces of actors and the bodies of dancers.
- Its physical structures work with those of the set, the blocking, the choreography.
- Its temporal structures to a greater or lesser extent control the rhythm of the piece. Its transitions are as important as its steady states – as Lucy Carter says: 'we all know, if you get into a cue in the wrong way the cue can just feel wrong, even if it's right'.
- It works in harmony with music and other sound, and any physical changes in the space (set changes, etc.).
- Like poetry, it pays special attention to its component parts *and* its overall structure – its form. It can be useful to think of the analogue of words as the quality of light and dark in each volume of space in each moment of the performance and the analogue of form is the cue structure.
- It has a consistent language throughout, which may make use of the conventions of theatre, of camera-based arts, of fine art, of digital art...
- What it does is more important than what it looks like.
- Affect on the audience is more important than any single technical effect.
- It is not afraid to be *the thing itself*; to show itself as light, beyond decoration and support for text, movement, or spatial and surface design.
- It has been made with emotion, passion, soul – call it what you will, but its author has risked, has exposed themselves. In short it has been made by an active practitioner – an artist.
- And it is reproduced each night by technically accomplished people who put their passion and emotion into its re-creation.
- It is purposeful. It has beauty and a quality of finish, and it is good at being the lighting design for that production, at each moment and throughout the whole piece.

It is right for that moment of performance. Not right for every production of that piece, and often not right for subsequent reproductions of the piece – theatre makers and audiences change over time and so does their understanding of what is the right light. But it is right for that moment of that production of that piece, for those performers and creative team members (which will usually include many people who are more usually called technicians) and for that performance space. It is right for the audience, helping them to see clearly what needs to be clearly seen, hiding that which it is useful to hide, heightening affect where appropriate and making a significant contribution to the experience of being an audience member for a piece of live performance. It is the right light.

Notes

1. Active Practice

1 For a fuller explanation of theatre's love affair with technology through the ages, see *Theatre, Performance and Technology* (Baugh, 2005 & 2013).
2 For a more complete history of the role of lighting designer up to 1950 in the UK see *Stage Lighting Design in Britain* (Morgan, 2005) and *Light*, chapters 1 and 12 (Palmer, 2013).
3 The artistic or creative team is generally assembled for a specific production – though there are exceptions where a team might create a season of work. One of the small and perhaps under-reported marvels of theatre is the way in which a team of creative individuals can meet for the first time for a production, collaborate over a period of weeks or months, and create a 'show' on time and in budget. It is interesting to contrast this with the many stumbles and delays in delivery typical of collaborative projects in other areas of endeavour.
It is perhaps worth noting that in almost all cases performers are not currently included in the creative team. The rights and wrongs of this and the implications it has are beyond the scope of this book, however.
4 See *Integrative Performance, Practice and Theory for the Interdisciplinary Performer* (Bryon, 2014).
5 For a detailed discussion on the influence of both men on contemporary theatre practice see (Palmer, 2013) and (Baugh, 2005 & 2013).
6 The original London Weekend Television series *Upstairs Downstairs* was first broadcast in the UK in 1971.
7 It should be said that there are clear exceptions to this generalisation. Michael Hulls works almost exclusively in contemporary dance – for reasons that Michael talks about later in the book.
8 Fischer-Lichte, 1992, reprinted in *Light* (Palmer, 2013).
9 See *Light* (Palmer, 2013) for a well-considered review of the use of light in stage design from Classical Greece onwards.
10 In an interview with psychiatrist Dr Anthony Clare for the BBC Radio programme 'In The Psychiatrist's Chair', in October 1998 (BBC, 1998).
11 DoP = Director of Photography, responsible for what the camera shoots and how it shoots it – where each shot is taken from and what is in the frame of each shot, and how it is lit. In other words, much of what the cinema audience see and how they see it.
12 *Analysing Performance* (Pavis, 2003 (published in the original French in 1996)).
13 One example of kind of normalisation I'm talking about is the way in which we quickly stop perceiving the colour shift in even heavily tinted sunglasses. If you want to do a little experiment, take a photo of a bright scene through a tinted filter such as the lens of a pair of sunglasses, and compare it to what you see through the same lens. A detailed discussion of the psycho-physics of human

vision is beyond the scope of this book. See for example *Vision and Art: The Biology of Seeing* by Margaret Livingstone (Livingstone, 2002). Although some of the science in this very accessible text is a little outdated now, it remains a good introduction to the subject.

14 *Focus* is the bi-monthly magazine of the British-based Association of Lighting Designers (ALD). It's strap line, established by a previous editor, Andy Collier, is: 'more art, less tools.'

15 To quote the Wordle website, www.wordle.net: 'Wordle is a toy for generating "word clouds" from text that you provide. The clouds give greater prominence to words that appear more frequently in the source text.'

2. Instinct as Inspiration

1 See (Palmer 2013) pp. 207ff for more on this.

2 *Our Lady of Sligo* by Sebastian Barry was an NT/Out of Joint production in 1998. Director, Max Stafford Clark; Designer, Julian McGowan; Lighting Designer, Johanna Town; Music, Corin Buckeridge; Sound Designer, Paul Arditti.

3 *Tonight at 8:30* by Noel Coward was an English Touring Theatre production in the spring of 2014. The piece consists of nine one-act plays, three of which are performed for each audience. Director, Blanche McIntyre; Designer, Robert Innes-Hopkins; Lighting Designer, Johanna Town; Sound and Music, Adrienne Quartly.

4 Lighting fixtures able to produce a wide range of colours. See Light Sources and Colour Mixing in the Glossary.

5 Earlier in the interview, Chivers talks about how she was inspired to change from stage management to lighting by seeing LD Simon Corder focusing in the Almeida theatre, and how he made the light from a single fixture transform the space.

6 The international hit theatre production, based on a novel by Michael Morpurgo, adapted by Nick Stafford, made in association with Handspring Puppet Company, premiered in 2007 on the Olivier stage of London's National Theatre. Director, Marianne Elliot and Tom Morris; Set and Costume Design, Rae Smith; Lighting Design, Paule Constable; Music and Sound, Adrian Sutton and John Tams.

7 *Danton's Death*, by Georg Büchner in a new version by Howard Brenton premiered on the Olivier stage of the National Theatre in July 2010. Director, Michael Grandage; Set and Costume Design, Christopher Oram; Lighting Design, Paule Constable; Sound and Music, Adam Cork. The curved back wall of the set was dominated by a large set of windows with shutters through which streamed large amounts of light.

8 Daily theatre or concert touring involves unloading and setting up the show, doing the show, breaking down and packing the show back onto the trucks, and then moving on – all in a single day. The first task each day for the touring chief electrician or lighting director is often to negotiate with the technical staff in the venue the positions of all the flown elements (the suspended elements of the lighting rig, any suspended speakers and masking), where the touring electrical distribution, dimmers and amplifiers, etc., will go, and many other details of how the show will fit into this venue today. Then she or he has breakfast, then through the day you learn how many of your decisions were good today. This is work that both Constable and I did earlier in our careers.

9 *The Light Princess*, a musical by Tori Amos and Samuel Adamson, premiered on the Lyttelton stage of the National Theatre in October 2013. Director, Marianne Elliot; Set and Costume Design, Rae Smith; Lighting Design, Paule Constable.

10 National Theatre's acclaimed production of *The Curious Incident of the Dog in the Night-Time* is based on Mark Haddon's award-winning novel. Adapted for the stage by Simon Stephens; Director, Marianne Elliott; Set and Costume Design, Bunny Christie; Lighting Design, Paule Constable; Music, Adrian Sutton; Video Design, Finn Ross.

11 Ormerod's work on this production was widely praised. David Allen wrote this for the *New York Times*: 'with powerful lighting by Ben Ormerod that uses clarifying chiaroscuro and block colors, the aesthetic is more obviously that of postwar New Bayreuth.' http://www.nytimes.com/2015/06/18/arts/music/review-tristan-und-isolde-brings-thunder-to-the-longborough-festival.html?_r=0.

12 Theatre Set and Costume designer – who also ran the Motley Design course for many years. Trained at the Central School of Art and Design at the same time as Peter Mumford.

13 When Macbeth sees the ghost of his old friend Banquo, whom he has recently had murdered.

14 Many stage lighting text books use the term 'key light' in the same way that someone writing about lighting for camera would – that is, the (apparently) single source that puts the main light into the scene and that light is for the faces of the performers (see Pilbrow 1997 et al). Poet is using the term to talk about what others have called 'motivational light' – something which often goes beyond the idea that it should first of all light the faces of the performers.

15 *Medea* at the National Theatre. Directed by Carrie Cracknell; Set & Costume Designer Tom Scutt; Lighting Designer, Lucy Carter; Music by Will Gregory and Alison Goldfrapp. The production opened in July 2014 and thanks to the NT Live initiative is also available on DVD.

3. Tech: A Cauldron of Potential

1 *Charlie and the Chocolate Factory* opened at the Theatre Royal Drury Lane in June 2013. Director, Sam Mendes; Set & Costume Design, Mark Thompson; Lighting Design, Paul Pyant; Projection Design, Jon Driscoll. Pyant and Driscoll jointly won the Olivier for Best Lighting Design, Thompson won the Olivier for Best Costume Design.

2 In stage lighting circles the words 'plot' and 'record' are more or less synonymous, except that 'plot' also refers to the schematic drawing of the lighting system, especially in North America. In the UK this drawing is more usually called the 'lighting plan.'

3 The practice of creating lighting looks on stage requires the lighting designer to continually access both their higher aesthetic functions and to master the recall of particular numbers and sets of numbers that are required to get the supporting technology to produce the look on stage. It would be interesting to monitor the brain activity of a lighting designer at work, alternately engaging different areas of the brain perhaps, or more likely over time developing then relying on very strong connections between different processing centres within the brain.

4 *Boeing Boeing* by Marc Camoletti at Sheffield Crucible Theatre, May 2014. Director, Jonathan Humphreys; Designer, Fabrice Serafino; Lighting Designer, Natasha Chivers; Music and Sound, Ben and Max Ringham.
5 A musical with music by the Gypsy Kings, opened at the Garrick Theatre in London's West End in July 2008. Director Christopher Renshaw; Set Designer Tom Piper; Lighting Designer Ben Ormerod.
6 Early lighting consoles or desks did not have a memory. Instead one or sometimes more operators would physically move dials, levers or faders to a pre-set mark for each new cue state.
7 Stories of mammoth overnight plotting sessions fuelled by alcohol and other substances, though once common in UK theatre, are now mostly a thing of the past. Most people who remember these overnight sessions admit that what resulted was rarely of a high standard. As a working practice it may also be at least partly responsible for several premature deaths.
8 The London revival of the Cameron Mackintosh musical *Miss Saigon* opened in the spring of 2014 at the Prince Edward Theatre. The creative team included: Book & Lyrics, Alain Boublil; Concept Book & Lyrics, Claude-Michel Schönberg; Director, Lawrence Connor; Set Design, Totie Driver, Matt Kinley; Original Set Design, Adrian Vaux; Costume Design, Andreane Neofitou; Lighting Design, Bruno Poet; Sound Design, Mick Potter; Musical Staging, Bob Avian; Additional Choreography, Geoffrey Garratt. More information at http://www.miss-saigon.com/creatives/ (accessed 9/9/14).
9 Mark Jonathan was head of lighting for the National Theatre in London. During that time, and occasionally since, he has taught master-classes in understanding lighting for theatre directors at the National Theatre Studio.
10 For example, in an article in the Continuum Companion to Twentieth Century Theatre: 'The lighting designer is "super-audience" and must develop an ability to sense when the dynamic is right, allowing the audience easy entrance to the piece, at the same time being totally involved in the mechanics and problems to overcome the actual process of lighting the play' (Chambers, 2006, p. 445).
11 Re-lighting is the job of reproducing the lighting design of a production on tour or when it comes back onto a repertory stage. It is an important stepping stone in the careers of most if not all of the lighting designers interviewed here, and most lighting designers working in theatre today.
12 A one-man adaptation of Shakespeare's *Macbeth* for National Theatre of Scotland and the Lincoln Centre Festival Directed by John Tiffany, with Set and Costume Design by Merle Hensel, performed by Alan Cummings, at the Tramway Glasgow in 2012 and Ethel Barrymore, Broadway New York in 2013.
13 'Block' is a type of cue state that is especially important when programming in 'tracking,' that is when changes made in the first cue state of a section 'track through' all the other cue states in that section. In lighting programming terms, a block is a class of cue state that ensures changes made to cue states before the block don't affect cue states after the block. For a detailed explanation of all aspects of programming lighting consoles, see Schiller, B. *The Automated Lighting Programmer's Handbook*, (Focal Press 2010).
14 This 'five star' production for Sheffield Crucible transferred to the Gielgud Theatre in London's West End in February 2005. It was directed by Michael Grandage with set and costume by Christopher Oram and a score by Adam Cork. Theatre critic Michael Billington wrote: 'Paule Constable's spooky lighting conjures shadows from which Velasquez-costumed figures emerge like

power-playing phantoms.' (Guardian newspaper 4/2/05. also at http://www.theguardian.com/stage/2005/feb/04/theatre).

15 This 'four star' production for the Donmar Theatre, again directed by Michael Grandage, with sets and costume by Peter McKintosh, premiered in June 2011. Michael Billington wrote: 'Paule Constable's lighting, filtering through the high windows of Peter McKintosh's black-walled set, is consistently ominous.' (Guardian newspaper 13/6/11. also at http://www.theguardian.com/stage/2011/jun/13/luise-miller-review)

16 The opera by Harrison (Harry) Birtwistle was premiered in 1986 by English National Opera, with elaborate staging including masks and large-scale puppets designed by Jocelyn Herbert.

4. Collaborations

1 Frantic Assembly was founded in 1994 by Scott Graham, Steven Hoggett and Vicki Middleton, to create 'thrilling, energetic and unforgettable theatre'.

2 *The Three Sisters* by Anton Chekhov: Director, Benedict Andrews; Design, Johannes Schütz; Costume, Victoria Behr; Light, James Farncombe; Sound Paul Arditti. The production premiered at the Young Vic on 13 September 2012.

3 *Bugsy Malone*, directed by Sean Holmes (who is also the artistic director of the Lyric Hammersmith) with set and costume designs by Jon Bausor, ran at the Lyric Hammersmith from April to September 2015 to largely four- and five-star reviews.

4 *A Doll's House*, directed by Peter Hall, set by Simon Higlett, costume by Christopher Woods, opened at Theatre Royal Bath in July 2008.

5 *La Bête* played at London's Comedy Theatre in 2010 and featured Mark Rylance as Valere. Director, Matthew Warchus; Set and Costume Design, Mark Thompson; Lighting Design, Hugh Vanstone.

6 *The Keepers of Infinite Space* by Omar El-Khairy with contributions from Caroline Rooney, at the Park Theatre London in January 2014. Director, Zoe Lafferty; Set Designer, Philip Lindley; Costume Designer, Susan Kulkarni; Lighting Design, Johanna Town; Music and Sound, Richard Hammarton; Dramaturge, Carissa Hope-Lynch.

7 The follow-spot positions at Drury Lane are over 100 feet or 30 metres from the stage!

8 *An Inspector Calls* by J.B. Priestley in a version directed by Stephen Daldry has been extensively revived since its first production for the National Theatre in 1992. Set and Costume Design by Ian MacNeil, Lighting Design by Rick Fisher, Incidental Music by Stephen Warbeck. It has won 19 major awards.

9 The opera *Wozzeck,* Directed by Keith Warner, Sets by Stefanos Lazaridis, Costume by Marie-Jeanne Lecca, made in 2003, when it won the Olivier Award for Best New Opera Production.

5. Dramaturgy: Light Telling Stories

1 *La Traviata* for Scottish Opera 2012/13. Director Annilese Muskimmon; Designer Nicky Shaw; Lighting Designer Mark Jonathan.

2 The National Theatre's adaptation of Mary Shelley's story was a 'sell-out hit' in 2011 and has subsequently been a hit broadcast for NT live. Adaptation by Nick Dear; Director, Danny Boyle; Set Design, Mark Tildesley; Costume Design, Suttirat Larlarb; Lighting Design, Bruno Poet; Music and Sound Score, Underworld.
3 *Versailles* was a new play by Peter Gill, produced for the Donmar Warehouse in February 2014, Directed by Peter Gill; Set and Costume by Richard Hudson; Lighting by Paul Pyant; Sound by Gregory Clarke; music consultant Terry Davis. The production ran to sell-out houses till April 2014 with four- and five-star reviews.

6. Dance and Abstraction

1 Made for Dancelines Film in 1988.
2 *Stop Quartet* was made in 1996. The piece was choreographed by Burrows, who also danced it with Henry Montes, Ragnild Olsen and Finn Walker to music by Kevin Volans and Matteo Fargion, and lighting design by Michael Hulls. The piece has become iconic in contemporary dance circles, and has been revived several times. In 2008 the dance critic Judith Mackrell wrote this about the piece: 'Part brain-teasing puzzle, part play, the Quartet used music ... to develop a complicated choreographic structure of humorous dance gesture and rhythmic invention' http://www.theguardian.com/stage/2008/oct/18/dance-londonlistings1 (4/8/14). The piece has been filmed and can be found on YouTube.
3 *Two* was originally created by Russell Maliphant for Dana Fouras in 1997 with Michael Hulls' lighting design and music by Andy Cowton. 'Michael Hulls's extraordinary lighting turns Guillem's hands and feet into licks of fire, while Maliphant's choreography engulfs her in its whirligig. The ballerina's body seems to evaporate into the vortex.' Quote from the advertising playbill of the Mariinsky Theatre for *PUSH*, in which Sylvie Guillem performed *Two*. October 2014.
4 *Afterlight (Part One)* was made for Sadler's Wells by Russell Maliphant and Michael Hulls and premiered in 2009. Music by Eric Satie; Costume Design by Stevie Stewart; Video Animation by Jan Urbanowski; Sound Design by Andy Cowton. Performed by Daniel Proietto. More information and images at http://www.russellmaliphant.com/work/afterlight/ (4/8/14).
5 John Tavener's suite *The Protecting Veil* was choreographed by David Bintley for Birmingham Royal Ballet, with set and costume design by Ruari Murchison, lighting design by Mark Jonathan. It premiered in 1998. Bill Cooper's image from the opening 'coup de théâtre' was extensively used by BRB, and the ALD for several years.
6 *Shift* commissioned by Dance 4, Nottingham and DanceXchange Birmingham, premièred in 1996. Choreographed and Performed by Russell Maliphant; Lighting Design, Michael Hulls; Music, Shirley Thompson; 'There is only one person on stage, the choreographer himself, yet it quickly becomes obvious that he is not dancing alone. For *Shift* is a virtual duet between Russell Maliphant the dancer and the ingenious lighting of his long-time collaborator, Michael Hulls. The interplay turns an intimate dance for one into a beguiling dialogue about

perception and the nature of reality.' Programme notes from the 2014 UK tour of *Push* credited to Debra Craine.

7 The Rodin Project is a co-production by Maliphant's company with Sadler's Wells Theatre, where it was first shown in February 2012. Choreography, Russell Maliphant; Lighting Design, Michael Hulls; Set Design, Es Devlin & Bronia Housman; Costume Design, Stevie Stewart; Dancers, Tommy Franzen, Thomasin Gülgeç, Ella Mesma, Carys Stanton & Jenny White. Images of the production can be seen at http://www.russellmaliphant.com/work/the-rodin-project/ (4/8/14).

7. In Search of the Right Light

1 The museum is in Gujarat in Western India. Their website is https://calicomuseum.org/.

2 See note in chapter 2.

3 The RSC production of *Wolf Hall*, from the novel by Hilary Mantel, adapted for the stage by Christopher Poulton, opened in Stratford in early 2014, directed by Jeremy Herrin, set and costume design by Christopher Oram with music by Stephen Warbeck. The production transferred to the Aldwych theatre in London in May 2014. Michael Billington wrote: 'Christopher Oram's design, Stephen Warbeck's music and Paule Constable's lighting concept, in which the world seems to brighten every time the king appears, are also beyond praise.' (http://www.theguardian.com/stage/2014/jan/09/wolf-hall-bring-bodies-swan).

4 *Two* was originally created by Russell Maliphant for Dana Fouras in 1997 with Michael Hulls' Lighting Design and Music by Andy Cowton. 'Michael Hulls's extraordinary lighting turns Guillem's hands and feet into licks of fire, while Maliphant's choreography engulfs her in its whirligig. The ballerina's body seems to evaporate into the vortex.' Quote from the advertising playbill of the Mariinsky Theatre for *PUSH*, in which Sylvie Guillem performed *Two*. October 2014.

5 *Shift* commissioned by Dance 4, Nottingham and DanceXchange Birmingham, premièred in 1996. Choreographed and Performed by Russell Maliphant; Lighting Design, Michael Hulls; Music, Shirley Thompson. 'There is only one person on stage, the choreographer himself, yet it quickly becomes obvious that he is not dancing alone. For Shift is a virtual duet between Russell Maliphant the dancer and the ingenious lighting of his long-time collaborator, Michael Hulls. The interplay turns an intimate dance for one into a beguiling dialogue about perception and the nature of reality.' Programme notes from the 2014 UK tour of *Push* credited to Debra Craine.

6 *Ghosts* by Henrik Ibsen was produced by the Almeida theatre in Autumn 2013 and went on to transfer to the Trafalgar theatre. It was highly acclaimed, and leading actor Leslie Manville won several awards for her performance. Director, Richard Eyre; Set and Costume Design, Tim Hatley; Lighting Design, Peter Mumford; Sound Design, John Leonard.

7 *Timon of Athens* at the National Theatre London in 2012. Director, Nicholas Hytner; Designer, Tim Hatley; Lighting Designer, Bruno Poet; Composer, Grant Olding; Sound Designer, Christopher Shutt.

8 For several nights in the summer of 2011 in various parts of London and other English cities, the streets were taken over by rioters, some of whom set fire to public and commercial buildings. The riots left five people dead and caused an estimated £200 million worth of damage.

9 I'm deliberately avoiding saying higher level of visual literacy. If we think about generations of ordinary people who could not read or write, but were able to read the images of, for example, stained glass windows, and take from them much more than anyone but a specialist could today, they can hardly be described as having poor visual literacy.

Glossary

Affect (noun) the experience of a feeling or emotion; in the theatrical context the emotions experienced by an audience as a result of what is happening on stage.

ALD Association of Lighting Designers.

Angles for a Lighting Designer

Conventionally, when performance lighting practitioners talk about angles, they refer to the angle of incidence to a performer facing straight out towards the centre of the audience. There are two angles involved: the angle of elevation, taken from the plane of the stage, which is assumed for these purposes to be flat, and the angle the performer would need to turn from looking out to centre, in order to face the lighting position. High or steep angles are close to directly overhead and low or shallow angles are close to or even below the eyeline of an upright performer. Backlight comes from behind the performer (180 degrees, or ½ a turn for the performer), side or cross light from the side (90 degrees or a ¼ turn in either direction) and top light from directly above. All of these terms can be modified or combined to describe different lighting positions relative to the subject or to the performance area. From Moran, *Performance Lighting Design*, 2007.

Lighting designers also use 'angle' to talk about how the light spreads out from a fixture – the beam angle. Small angle fixtures have a narrow beam while fixtures with large beam angles have a wide beam that quickly spreads out.

Associate/Assistant

These are roles that help the lighting designer fulfil their role. The associate deputises for the lighting designer, and will often be charged with reproducing the design. The assistant will usually take notes, and look after paperwork (and often the refreshment and other needs of the lighting designer), but should not make independent decisions, even in the absence of the lighting designer. Both roles can also provide a second pair of eyes and someone to bounce ideas off.

Backlight

Generally, light from behind the performer's head, when the performer is facing the audience. It can have a significant architectural function (backlight beams are more visible to the audience) as well as helping to reveal forms on stage in a more sculptural and three-dimensional way.

Backlight helps to separate the performer's body and other foreground elements from the scenic background. As Johanna Town says, 'you take it out and suddenly everybody goes flat, even though you don't actually think you can see it.'

Birdie

A miniature lighting fixture, which may be concealed in the set or fixed to the front edge of the stage to simulate footlights.

Blocking

The positions of actors on the stage and the paths they take to arrive there; also the process of deciding this. Blocking may be tightly defined (by the director or by technical considerations) or entirely free (as when the performer is able to decide in the moment where on the stage they will be at any given moment and what route they will take to get there). In practice, most performers in most work settle into moves that they stick to more or less, for every performance. This makes it more possible for the LD to (for example) create lighting states that work in a consistent way from one performance to the next. Completely free blocking usually requires the LD to respond in one of three ways: first, use a follow-spot; second, flood the performance area with light (often sacrificing other aspects of the design in the process); or accept that there will be times when the performer is not lit in quite the right way. As Peter Mumford says in chapter 4: 'for the most part, actors like to have a precise track through a piece, they like to know where the other actors are going to be, and the kinds of nuances they'll put on it will be largely internal, it won't be about, "ok, now I'm actually going to do this from the other side of the room". So you can then work with them, I think, and teach them to love the light.'

The Book (i.e. DSM Script)

This is a marked-up copy of the script used by the DSM or show caller to call the show. Ideally it contains all the information required to run the show, all the positions of all the cue-points, as well as blocking and pacing notes for the cast, etc.

Booms

Vertical lighting positions, usually at the sides of stage in the wings, though they may be anywhere in the performance space. Side booms are the main lighting position for much dance lighting.

BP Screen

Literally, back projection screen, but often used to describe any large seamless plastic drop used as a cyclorama.

Brightness see **Intensity, Contrast and Brightness, for the eye and for camera**

Cans, aka Headsets or Talkback

Combined headphones and a microphone used to provide two-way communication with the lighting programmer, stage management team and others backstage. Usually wired, they can now be wireless, which can mean the lighting designer has more physical freedom to see the stage from places other than behind the production desk.

Channel Number

The number used by the lighting designer and programmer to control a lighting fixture.

Cheated in/out or on

1. To very slowly raise or lower the intensity of one or more units usually to correct a look without drawing the attention of the audience away from where it needs to be focused.
2. To provide just enough light to fill the shadows created by a strong key source, without drawing attention to the new source.

Chief Electrician/Production Electrician

The lead lighting crew person, directly responsible to the lighting designer for getting the rig up and working. Generally, the chief electrician is a full-time employee of the theatre, while the production electrician is a freelance contractor of the production company, just as most lighting designers are.

Colour Change Technology and Colour Fades

Theatre has a long history of finding ways of changing the colour of light coming from a single fixture (see Palmer, 2013). Current technology means this can now be done smoothly and continuously – the colour fade from one colour to another.

This is an important enabling technology for many lighting designers since it allows them to change the colour of a shadow or whatever, with a single unit rather than fading one unit out while another unit fades in, which moves the shadows.

Colour Numbers

To colour the light from most conventional fixtures, lighting designers use coloured plastic filters. This is frequently called 'gel sort' for gelatine – a substance not used in professional theatre for several generations, but some things change only slowly. Colour filter comes in sheets and rolls; each range has swatch-books, and each different tint and hue has a name and a code number. Most of the colour ranges are now also catalogued on apps. The main ranges used in UK theatre are Lee and Rosco. Both companies have websites where you can see the hundreds of different filters available.

Knowing the code number of many of the frequently used colours is an expected part of the vocabulary for theatre lighting people, though this may change as conventional fixtures are replaced by ones that have internal ways of changing the colour.

Contrast and Brightness see **Intensity, Contrast and Brightness, for the eye and for camera**

Cool and Warm Light

These are terms that relate the perceived colour temperature of light.

Warm: as in the comforting light of a domestic fire, the glow of a sunset in fine weather, the atmosphere between contented lovers or amongst a happy family or group of friends. In performance lighting terms: orange, red and yellow tints, perhaps relying on incandescent sources at lower intensity to increase the proportion of red light in the 'white'.

Cool: as in the light of a clear bright winter day, the light of clinical or scientific spaces, the atmosphere between recently separated lovers, or amongst a family split by arguments. In performance lighting terms: harsh and revealing, lots of pale blue tints and no hints of reds or orange, perhaps making use of fluorescent and white discharge sources, or incandescent lamps at full. From Moran, *Performance Lighting Design*, 2007.

Cross-Fade see **Fade, Cross-Fade, Snap**

Cue/Cue Point/Cue State

In general, the moment in time during a performance when a change starts. This might be a change in which character is speaking, the start of a move (of e.g. performers or stage machinery) or the start of a change in the lighting. This is generally the meaning when someone says, 'where does this cue go?' For added clarity I have used 'cue point' for this meaning.

It is also used by lighting people and others as shorthand for 'cue state', and this is generally the meaning when someone says 'this cue is too fast' or 'that cue was too bright downstage'.

The problem comes when someone says something like, 'that cue was wrong!' – was the cue point-wrong (e.g. called in the wrong place) or the cue state wrong (e.g. too bright downstage)?

Cue List

A cue list for a lighting designer is a list of all the cue states, with some additional information, which may include some or all of the following:

Cue position; motivation for the change and/or required affect of the cue state; details of times; details of key sources used; other notes.

For more detail see Moran, *Performance Lighting Design*, 2007.

Cue Time

The time taken to execute a lighting cue (or many other kinds of cue for that matter.) Cue times can be simple (every change happening over the same period of time) but more often than not major changes in the lighting on stage will involve multiple times, some for aesthetic reasons and some for technical ones.

There is more on this in chapter 3 and in Moran, *Performance Lighting Design*, 2007.

Cyclorama or Cyc

The cyclorama used to be the smoothly plastered, white painted curved back wall of the stage, lit to create the illusion of infinite depth. These days, however, it is any large plain illuminable backdrop. The word is frequently shortened to cyc and may be used interchangeably with BP in some circumstances.

Discharge Lamps see **Light Sources**

Dress Rehearsals

Usually the final rehearsal before the show is put in front of an audience (though in some institutions dress rehearsals are open, that is, there may be public there). It is not usually possible to stop a dress rehearsal to make adjustments to anything, including lighting; however it is usually acceptable to make subtle changes to lighting cue states and times.

DSM (deputy stage manager)/Show Caller

Uses the book to call the show – that is to let all those who need to know when to do what they need to do. Most of what a show caller does could be described as 'following the script and calling out cues to operators as each cue point is reached'.

However, there can be a lot more to it than that (see Moran, *Performance Lighting Design*, 2007).

During the technical rehearsals, the DSM/show caller and the programmer are generally in constant communication with the lighting designer.

Drop

A name for a flown piece of scenery, usually cloth with a stiffening pipe in the bottom, wide and high enough to fill the stage, masking whatever is behind (or upstage). The term can be further modified e.g. 'backdrop' – a piece of scenery which creates the background of one or more scenes – and 'show-drop'; flown downstage as an alternative to the stage curtains, painted with a show logo or scene-setting image. Drops usually require specific instruments to light them, and getting the light right on the drops can be a difficult and time-consuming business.

Fade, Cross-Fade, Snap

Ways of describing different timings for transitions between cue states. See also Chapter 3, Transitions, Cue Times and Dynamic Looks.

Fill Light see **Key Light and Fill**

Fixtures see **Instruments/Fixtures/Lanterns/Lighting Rig**

Fly Cue

1. A timed change in the position of suspended set pieces or other suspended elements (which can include parts of the lighting rig – see Mark Jonathan on *The Protecting Veil* in chapter 6). Flying something in means lowering it while flying something out means raising it, generally into the void above the stage where it will be out of sight of the audience.
2. The point in time when the change starts (see also Cue/Cue Point/Cue State).

Focus

1. The process of pointing and otherwise adjusting the fixtures of the lighting rig. It is usually directed by the lighting designer and carried out by the chief electrician and her or his crew. Traditionally it required some 'bravery' at height (which might also be characterised as naive failure to acknowledge risk), but with today's increased focus on the health and safety of theatre workers, the exposure to risk is significantly reduced. This has been achieved through the introduction of safer ways of working at height, and moving light technology, which means these fixtures are focused from the lighting console.
2. The point of maximum attention (on stage). Usually the brightest area in the visual field (of the audience).

Follow-Spot/Follow-Spot Operator

Often abbreviated to 'spot'. Literally a spotlight operated by a person, the beam of which follows a performer (or anything else) round the stage, and sometimes beyond the stage. As well as being a way of highlighting one or more key performers – by lifting their visual intensity above that of their surroundings – it is also a signifier of star status in many genres of theatre and performance. This is especially so when the edge of the beam of light from the follow-spot is sharp. However, many lighting designers employ soft-edged follow-spot beams that are only just brighter than the surrounding illumination to subtly guide the attention of the audience. Good follow-spot operators are highly valued by the lighting designers who use them, though generally this is not reflected in their wages. (See Paul Pyant and David Howe in chapter 4, and Lucy Carter in chapter 6.).

Front of House

Frequently abbreviated to FoH, the term generally refers to the parts of the venue on the audience side of the proscenium or other boundary between performers and audience (such as the edges of a thrust stage closest to the audience). It is also used as shorthand for fixtures rigged on the audience side of the boundary. Often a theatre lighting rig will be organised into front of house (FoH) and on stage. Most of the lighting designers here make far less use of light from FoH fixtures than lighting designers following a more traditional practice, though there are exceptions.

General Cover

A way of describing the even washes of light that were a key element of traditional lighting practice, as described e.g. in McCandless, 1932, revised edition 1939 and Pilbrow, 1997. A general cover provides adequate illumination over the required performance area.

Gobo

Most often, a thin stainless steel plate with an etched through pattern. When the gobo is placed correctly in the correct type of fixture, the pattern etched into the gobo appears in the beam of light. Gobos can also be made in glass, when the image can then include colour. Many moving lights are able to hold multiple gobos. Lighting designers and programmers can use them to animate light in a number of different ways. To my knowledge there is no agreement on the origin of the word.

God Mic

A microphone on the production desk that allows the speaker to address the whole theatre. Traditionally used by the director during the tech to maintain his or her authority through asymmetrical communication (no-one else can address everyone etc. – hence 'God' mic), in more collaborative environments its use is confined to making sure everyone knows what is going on.

Headsets see **Cans, aka Headsets or Talkback**

HMI

Usually shorthand for a high intensity luminaire, a term originating in film lighting. They produce a quality of light similar in many ways to a shaft of daylight, though they are often used from lower angles than you would normally expect such strong sunlight shafts to come from. Very useful for mimicking a *film noir* look on stage.

Instruments/Fixtures/Lanterns/Lighting Rig

The three words instruments/fixtures/lanterns are used more or less interchangeably in theatre for the bits of equipment that the light comes out of.

The lighting rig is the assemblage of lighting fixtures in the correct places, with all the associated technological and physical infrastructure required to get them into those place safely, with appropriate power, and under the control of one or more lighting consoles.

Intensity, Contrast and Brightness, for the eye and for camera

For lighting designer and other lighting people, 'intensity' most often refers to the amount of light coming from a fixture. When the lighting designer is working with a programmer, she will most often be making adjustments in the intensity of each fixture involved in the look. In this context, the intensity is referred to as a percentage of the fixture's full intensity. It is an objective measurable quantity, recorded in the lighting console for each fixture in each look.

When lighting designers use the word 'contrast' they are generally referring to how bright or dark the foreground is in comparison to the background – often then it is a measure of the visual dominance of subject over ground (and occasionally vice versa). This is an objective measurable ratio, though the perception of contrast is subjective in humans.

When they use the word 'brightness', it may be in relation to part of the stage picture, the whole stage picture or a distraction outside the stage picture. Brightness is linked to but not the same as intensity. In terms of human perception, brightness is a relative term. Something that looks bright in an otherwise dark theatre will not look bright in daylight, for example. Brightness also depends on what you have recently seen, and on what else you can see in your visual field – thus it is possible to make a subject on stage appear brighter by reducing the brightness of its surroundings, or by reducing the overall intensity of the fixtures used in the previous cue state.

Cameras do not deal well with extremes of brightness since they cannot 'see' as much detail in higher contrast pictures as humans. This is one of the issues frequently encountered when live performance is captured on camera; see Moran, *Electric Shadows*, 2013.

Key Light and Fill

The expression 'key light' comes primarily from lighting for camera, where the viewpoint is more defined that it is in theatre and live performance generally. However, the key light for a stage scene is generally the most important source for that look. When the lighting design for a scene is motivated by naturalism, the key may be light through a window, or light from a practical, or some other apparent source. When the approach is more abstract the motivation for the key light may be abstract too, or may be driven by other aesthetic or practical considerations.

The lighting designer will often begin developing the look by introducing the fixture or fixtures that provide the key light. This will then be balanced by the introduction of light that will fill the deep shadows created by the key, without destroying them and consequently flattening out the stage picture.

An approach based on motivated key light sources and fill light can be seen as an alternative to establishing a more uniform and less motivated general cover.

Lamps see **Light Sources**

Lantern Rig see **Instruments/Fixtures/Lanterns/Lighting Rig**

LED see **Light Sources**

Lighting Over

Lighting over is shorthand for the practice of creating the looks for each lighting cue state during the tech as opposed to in a dedicated lighting session beforehand. It has become an important part of the practice of all those interviewed here – led by the desire to light the bodies in the space rather than simply the space.

Lighting Plan/Lighting Plot/Light Plot see **Rig Plan/Lighting Plan/Lighting Plot/Light Plot**

Lighting Rig see **Instruments/Fixtures/Lanterns/Lighting Rig**

Lighting Session

Any time in the technical rehearsal period (and sometimes after it) when the work of the lighting designer is prioritised. This can include sessions for pre-programming moving lights or complex sequences, doing lighting notes or making adjustments to the lighting rig (such as re-colouring or re-rigging fixtures).

It is not unknown for the lighting designer to perceive time that other perceive to have a different priority, as a lighting session.

Light Sources/Lamps (including LED, tungsten-halogen and discharge lamps)

Until the mid to late 1980s, almost all light on British stages came from incandescent tungsten or tungsten-halogen lamps. On many stages these light sources still dominate, and are the favoured source of light for most of the lighting designers here most of the time. In these lightbulbs the light comes from a glowing tungsten filament. It has a spectral content similar to that of sunlight, and it fades in and out in a very smooth and gentle way, enabling looks to develop out of blackness with no distraction – something other technologies of light production have struggled to do.

However, things are changing fast. First with developments in so-called discharge lamps (lightbulbs where the light is produced by a controlled spark) and now with LED technology, moving lights and static fixtures of increasingly good quality are being produced, and the best of them can now grow a look out of a blackout as well as a conventional tungsten lighting rig. These new fixtures can be programmed to fade colour in what is perceived to be a natural way too (see also Colour Change Technology). It may be that the older incandescent tungsten technology of what are sometimes called conventional fixtures may be largely replaced in the near future.

Masking for the Stage

Various different ways of hiding what we don't want the audience to see, including elements of theatre technology that might cause unnecessary distraction to the audience – e.g. the lighting rig. Masking is also used to create entrances and exits from the performance area, hiding from view performers and items of set that are about to come onto stage or that have just left it.

Masking can be in almost any material and any colour. When it is made from suspended material it is called soft masking and that is often made from black wool serge curtains, and thought of as neutral masking by many theatre makers. The two most common forms of neutral soft masking are:

Italian masking: Legs & Boarders used to create a set of frames parallel to the front edge of stage. Legs create the sides of the frame, boarders the top. Upstage of each masking frame is space for a lighting boom.

German Masking: or Up – Down masking, creates a continuous box of masking material around the stage this scheme usually designed to create the impression of a limitless void beyond the performance area, as long as no light directly hits it!

Model Box

A scale model of the stage (and sometimes the whole performance space) used to evolve and then demonstrate the physical design. Creating the model box is usually the responsibility of the set designer. The scale used most often in the UK is 1:25.

The model box can be used by the director to help her work out the spacing and movement (blocking) of performers and objects, by the production manager and scenic constructors as a guide to pricing and building the set, and by the LD to try out angles and colour choices. It is often used to present the physical design of the production to the cast on the first day of rehearsals. Exhibitions of theatre design have traditionally focused on displaying the model box, which has often been tidied up or even completely re-made for that purpose. (See also white card model.)

On Stage/Offstage & Upstage/Downstage

On a proscenium stage, on and off refer to sideways movements of someone facing the audience, on stage being a move towards the centre and offstage being a move towards the sides or wings. Upstage and downstage refer to moving towards and away from the audience respectively – this terminology comes from raked stages, which have a slight slope down towards the audience that helps sight-lines for a flat auditorium. Problems arise when these terms are applied on any stage without a proscenium.

PAR

Acronym for parallel aluminised reflector and the shorthand name for the PAR 64 lighting fixture. This is one of the workhorses of performance lighting, a simple static tungsten halogen unit with a powerful beam that (signally or in clusters) does a great job of (for example) imitating sunlight through windows.

Pipe-End Wash

A way of getting side-light onto a proscenium stage from the overhead lighting rig. This is seen by many lighting designers as a very useful position, especially when little or no light can be squeezed in from the sides.

Pixel Mapping

A way of controlling individual elements of a large array of fixtures by 'mapping' the location of each element to a specific pixel in a video array. The virtual pixel in the array can be made to respond to the brightness and/or colour of the pixel it is mapped to in a video image. For *Frankenstein* (mentioned by Bruno Poet in chapter 5) the intensity of each small cluster of lightbulbs is determined moment by moment by the intensity of a specific pixel in the video image at that moment. This was a very effective way of creating dynamic waves of intensity across the array of lightbulbs.

Practical

This could be a working wall light or suspended chandelier in a realistic room set, a fire or candles, a torch, or something quite abstract such as a glowing cube.

Pre-Programming

Entering data into the lighting console ahead of the tech. This could be the empty cue list (see Jon Clark in chapter 3) or building blocks for cue states, or whole looks developed using (for example) visualisation software.

Production Desk

During the technical rehearsals, the main workplace of the lighting designer, and others. It can be anything from a board balanced across seats in the auditorium to a custom-made workplace with computer monitors, power outlets and dimmable work-lights. From the production desk, the lighting designer should be able to talk to the DSM, lighting programmer, follow-spot operators and others over cans, and with the director and other members of the creative team face to face. She should also have a good clear view of the stage.

Almost all the lighting designer's work of creating the lighting cues is done from the production desk.

Programmer

Sometimes referred to as the human to computer interface for the lighting designer this specialist – usually a member of the lighting crew – is responsible for entering data into the lighting console in order to produce a set of cue states and transitions. As lighting design has become increasingly dependent on moving lights and other technology, this role has become increasingly important. The moving lights, and the consoles, have become increasingly sophisticated and now offer many ways of achieving *almost* the same lighting state or transition. It often falls to the programmer to help the designer make choices that have both technical and aesthetic consequences. As can be inferred from the interviews, many LDs have strong working relationship with particular programmers, some of whom are considered to be associate LDs.

Proscenium Arch

In a conventional playhouse or opera house, the structure that frames the stage and by doing so creates hidden spaces above (the flys) and to the sides (the wings) of the stage.

Repertory

Any system of production that involves alternating productions in the same performance space, either daily (as is often the case with opera and ballet companies) or with longer runs of a few days or several weeks (as is the case on the stages of London's National Theatre, for example. Since the lighting designer is only paid to be with the production till opening night, each time the show is put back the lighting designer relies on the integrity and dedication of the in-house lighting crew to reproduce her lighting design.

In the opera and ballet world, repertory has another implication for the lighting designer since most large opera and dance companies play a mix of new productions and revivals of previous work in each season. This entails the re-production of work made anything from six months to many years previously. The subject of re-presentation of work in this way is touched on by international opera director Dr Jonathan Miller in his book *Subsequent Performances* (Faber & Faber, London 1988) and by Rick Fisher at the end of chapter 4.

Re-Lighting

Reproducing the work of another lighting designer, either on tour or in a repertory situation. Frequently acknowledged as a great training ground for emerging lighting designers.

Rig Plan/Lighting Plan/Lighting Plot/Light Plot

All terms that can mean the drawing (digital or ink on paper) representing the positions of fixtures in the lighting rig.

In addition, light plot can mean the list of cue states including their contents and timings, as recorded in the lighting console.

Run-through

A rehearsal where the aim is to do a section of the piece without stopping. (This is in contrast to a stopping rehearsal.) A full run-through is a rehearsal of the whole piece.

Site-specific/Site-responsive

Generally, theatre made away from a conventional theatre space, which in some way references the physical space and/or its history, though this referencing may not always be obvious. Away from the electrical and physical infrastructure of a theatre building, lighting design for site work (as it is often called) can present considerable practical as well as aesthetic challenges.

Sharp/Soft

Terms used to describe the quality of the edge of a beam or a shadow, sharp being well defined and soft less so. Both qualities have several degrees within them. The human eye is much more sensitive to rapid or sharp changes in intensity (rapid in space and in time) and so the lighting designer can make things more or less noticeable by varying the sharpness/softness of changes.

Show Caller see **DSM**

Snap see **Fade, Cross-Fade, Snap**

Specials

Usually single fixtures with a special purpose, such a highlighting a solo performer. Specials can be fixed fixtures or a particular position for a moving light. With planning, fixtures that work together to form a larger wash can also be used as specials individually.

Talkback see **Cans, aka Headsets or Talkback**

Technical Rehearsal/Tech

The start of the tech is the point when the performers and the director leave the rehearsal room and come onto the stage, in the set, under the lights, with costume and sound and everything else. It is the time when the show is put together technically. It is the period where each cue point for every performer and each technical operator is tested and defined. It is also the period when designers begin to find out if their ideas actually work. The tech is the part of the production process that most lighting designers find most intense. For a relatively simple play it may last less than a day. For a complex musical it may last several weeks. The tech for the West End production of *The Lord of the Rings* lasted several months.

Time-code

Comes from the film industry where it is used to synchronise image and sound. In live theatre and performance it can be used to synchronise a backing track to live vocals or performance – so it is often present on show backing tracks. (Not everything in live theatre is strictly live.) Modern lighting consoles can be set up to perform their cues to time-code, meaning that as long as the show backing track runs correctly, the lighting cues will be in the correct places every night.

Tracking

1. A way of thinking about and recording lighting cues for a show that can make lighting over much more possible. See chapter 9 of Moran, *Performance Lighting Design*, 2007 for more detail.
2. Using various technologies to enable moving lights to perform like automatic follow-spots and follow a performer. None of the systems so far developed is perfect, but some impressive results have been achieved.

Tungsten-halogen Lamps see **Light Sources**

Vari*Lite™

One of the first commercially available touring moving light systems, the trade-name has to some extent become synonymous with moving lights, in the same way that all vacuum cleaners tend to get called Hoovers. However, not all moving lights are Vari*Lites.

Visual Field

The entire area that can be seen at the same moment during a steady fixed gaze. For humans, this includes the well-focused centre of attention to the much less well defined peripheral vision.

Visualisation Suite

Digital pre-visualisation of stage lighting design is an extremely widely used tool in concert and event lighting, frequently enabling quite complex shows to be created with the minimum of live rehearsal. Now large opera and dance companies are beginning to introduce pre-visualisation, hoping that it will offer the same advantages gained in concert and event production to their process. For most large concert productions, however, spectacle is more important than the selective visibility of human faces and bodies. Theatre and dance lighting designers largely remain to be convinced that the software can do for them what it does for concert and event lighting designers. As Lucy Carter says, 'I can see when you're doing a massive rock show or something how brilliant it would be, but something that's much more subtle, I just don't think it's a reliable tool for that yet.'

Warm Light see **Cool and Warm Light**

Wash/Wash Light

1. Usually, a wash of light has a soft edge and appears to come from a single source (though may come from many fixtures focused in the same direction).
2. A wash light is a term used for a moving light that can produce a soft edge. There has been a general expectation for some years now that wash lights will smoothly fade between colours too.

White Card Model/White Card

The term is used interchangeably for both the outline or preliminary scale model (traditionally made in white card or some other cheap and easy to work material) and the meeting at which this model is presented, for costing and other feasibility discussions. In the traditional production process it was unusual for the lighting designer to see the white card model. As Ben Ormerod says, 'Sometimes, you're actually not in the country when the set's being designed. Sometimes you don't get booked until after the set's been designed, amazingly enough.' However, these days more and more production teams want the lighting designer to be involved from at least this stage.

The next stage in the production process is the presentation by the design team (including the director) of a fully realised model of the production design, usually at 1:25, for the makers and others to cost. If the lighting designer is not involved before this point, she or he is unlikely to be able to have any significant influence on the physical structure, which may severely constrain what can and cannot be achieved with light (see Constable in chapter 3).

Working Light/Workers

'Workers' is a frequently used shorthand for working light; that is, any lighting fixtures rigged and used purely as functional illumination and/or to make the space safe for technical work. If anyone is inclined to disregard the contribution the lighting department makes to a production, they are frequently shown the stage set lit only by the workers.

A Brief Bibliography

Barthes, R. (1980). *Camera Lucida.* (R. Howard, Trans.) Hill and Wang.

Baugh, C. (2005 & 2013). *Theatre Performance and Technology: The Development of Scenography in the Twentieth Century.* London: Palgrave Macmillan.

Belsey, C. (2002). *Critical Practice (2nd edition).* London: Routledge.

Brown, I. (2014, April 14). *Lighting Designer Michael Hulls.* Retrieved from the arts desk.com: http://www.theartsdesk.com/dance/theartsdesk-qa-lighting-designer-michael-hulls.

Burnett, K. (2013). *World Stage Design 2013.* OISTAT.

Bryon, E. (2014). *Integrative Performance, Practice and Theory for the Interdisciplinary Performer.* London: Routledge.

Chambers, E. b. (2006). *Continuum Companion to Twentieth Century Theatre.* London: A & C Black.

Deleuze, G. (1981 / 2003). *Francis Bacon: The Logic of Sensation (trans. from French Daniel W Smith).* London: Contunuum.

Mackrell, J. (2014, Feb 4). *Guardian Newspaper.*

McCandless, S. (1932 Revised edition 1939). *A Method of Lighting for the Stage.* New York: Theatre Arts Inc.

Miller, D. J. (1986). *Subsequent Performances.* London: Faber & Faber.

Moran, N. (2007). *Performance Lighting Design.* London: A&C Black.

Ost, G. (1956, second impression 1957). *Stage Lighting, a practical handboook.* London: Herbert Jenkins Ltd.

Palmer, S. (2013). *Light. Readings in Theatre Practice.* London: Palgrave Macmillan.

Pilbrow, R. (1997). *Stage Lighting Design, The Art, The Craft, The Life.* London: Nick Hern Books.

Rudge, H. C. (1925). *Stage Lighting for 'Little' Theatres.* Cambridge: W. Heffer & Sons.

Shepherd, S. (2012). *Direction. Readings in Theatre Practice.* London: Palgrave Macmillan.

States, B.O. (1985). *Great Reckonings in Little Rooms.* University of California Press.

Wertenbaker, J.R. (1972). *The Magic of Light: The Life and Work of Jean Roensthal, Pioneer in Lighting.* Little, Brown & Company.

Williams, R. (1976). *Keywords. A Vocabulary of Culture and Society.* Oxford: OUP.

Index